De Gruyter Proceedings in Mathematics

Monte Carlo Methods and Applications

Proceedings of the 8th IMACS Seminar on Monte Carlo Methods, August 29 – September 2, 2011, Borovets, Bulgaria

edited by
Karl K. Sabelfeld
Ivan Dimov

De Gruyter

Mathematics Subject Classification 2010: 65c05, 65c20, 11k40.
Physics and Astronomy Classification Scheme 2010: 02.70Uu, 05.10Ln, 87.10.Rt.

ISBN 978-3-11-029347-0
e-ISBN 978-3-11-029358-6

Library of Congress Cataloging-in-Publication Data

A CIP catalog record for this book has been applied for at the Library of Congress.

Bibliographic information published by the Deutsche Nationalbibliothek

The Deutsche Nationalbibliothek lists this publication in the Deutsche Nationalbibliografie; detailed bibliographic data are available in the internet at http://dnb.dnb.de.

Typesetting: PTP-Berlin Protago-TEX-Production GmbH, www.ptp-berlin.eu
Printing and binding: Hubert & Co. GmbH & Co. KG, Göttingen
∞ Printed on acid-free paper

Printed in Germany

www.degruyter.com

Preface

Welcome to the special issue of the Monte Carlo Methods and Applications journal with the selected papers presented at the 8th IMACS Seminar on Monte Carlo Methods hosted by the Institute of Information and Communication Technologies, Bulgarian Academy of Sciences in cooperation with the International Association for Mathematics and Computers in Simulation (IMACS). As mentioned in the previous special issue of MCMA presenting the 7th IMACS Seminar, over time, the IMACS Seminar on Monte Carlo Methods has created a history going back to April 1997, when the first MC Seminar was organized by R. Beauwens, A. Dubus, and M. Hou in Brussels. At this first IMACS MC Seminar, three invited plenary lectures were given (M. H. Kalos, K. K. Sabelfeld, and I. M. Sobol). The overview of the first eight IMACS seminars and the forthcoming 9th IMACS MC Seminar which will be organized by C. Lecot (Annecy, France) in 2013 is given in the following table:

# IMACS Seminar	*Date*	*Place*	*Organizers*
I	April 1–3, 1997	Brussels, Belgium	R. Beauwens A. Dubus, M. Hou
II	June 7–11, 1999	Varna, Bulgaria	I. Dimov
III	Sept. 10–14, 2001	Salzburg, Austria	K. Entacher W. Schmid, A. Uhl
IV	Sept. 15–19, 2003	Berlin, Germany	K. Sabelfeld
V	May 16–20, 2005	Tallahassee, USA	M. Mascagni
VI	June 18–21, 2007	Reading, UK	V. Alexandrov
VII	Sept. 6–11, 2009	Brussels, Belgium	A. Dubus, K. Sabelfeld
VIII	Aug. 29–Sept. 2, 2011	Borovets, Bulgaria	I. Dimov
IX	Aug. 26–30, 2013	Annecy, France	C. Lécot

We are deeply indebted to our colleagues who supported us in organizing the 8th IMACS Monte Carlo Seminar in Borovets. We also thank the reviewers for their very hard work in making this special issue possible. We all greatly appreciated the high quality of papers that we received from such a variety of contributors.

Novosibirsk, Sofia
July 2012

Karl K. Sabelfeld
Ivan Dimov

Contents

Chapter 1

Improvement of Multi-population Genetic Algorithms Convergence Time

Maria Angelova and Tania Pencheva

Abstract. Different kinds of multi-population genetic algorithms have been investigated for a parameter identification of a fermentation process. Altogether six realizations of multi-population genetic algorithms have been proposed, four of them with a different sequence of implementation of main genetic operators selection, crossover and mutation, and another two without mutation. A comparison of the considered six kinds of genetic algorithms is presented for a parameter identification of a fed-batch cultivation of *S. cerevisiae.* The influence of the most important genetic algorithm parameters, namely generation gap and rates of crossover, mutation, insertion, and migration, have also been investigated. All kinds of considered multi-population genetic algorithms lead to similar values of the optimization criterion. Among those with three genetic operators, the algorithm with a sequence of selection, crossover, and mutation is significantly faster than the others. When mutation is avoided, the genetic algorithm with a sequence of selection and crossover is faster than the other one.

Keywords. Multi-population Genetic Algorithms, Parameter Identification, Fermentation Process, Saccharomyces Cerevisiae.

Mathematics Subject Classification 2010. 37N25, 49J15, 90C31, 97M60.

1.1 Introduction

Investigation of fermentation processes (FP) is a question of present interest due to the fact that different branches of industry use them in the production of pharmaceuticals, chemicals and enzymes, yeast, foods, and beverages. Taking into account that FP are complex, nonlinear, dynamic systems with interdependent and time-varying process variables, their examining as modelling and control objects can present a serious challenge. An important step for adequate modeling of FP is the choice of a certain optimization procedure for model parameter identification. Genetic algorithms (GA) are a quite promising stochastic global optimization method which is widely applied for solving different complicated problems [2–5, 7–9]. GA, inspired by Darwin's theory of "survival of the fittest." [3], are one of the methods based on biological evolution. Properties like hard problems solving, noise tolerance, easy to interface and hybridize make genetic algorithms suitable and more workable for a parameter identification of fermentation models [1, 2, 4, 5, 7–9].

The basic simple genetic algorithm (SGA) initially introduced in [3] searches for a global optimal solution using three main genetic operators in a sequence selection, crossover, and mutation. SGA works with a population of the coded parameter set called "chromosomes". Each of these artificial chromosomes is composed of binary strings (or genes) of certain length (number of binary digits). Each gene contains information for the corresponding parameter. Multi-population genetic algorithm (MpGA) is a single population genetic algorithm in which many populations, called subpopulations, evolve independently from each other for a certain number of generations. After a certain number of generations (isolation time), a number of individuals are distributed between the subpopulations. According to [3], the structure of MpGA can be presented briefly as follows:

Algorithm 1.1.

Begin
1. [**Start**] Generate k random subpopulations each of them with n chromosomes
2. [**Object function**] Evaluate the object function of each chromosome n in the subpopulations
3. [**Fitness function**] Evaluate the fitness function of each chromosome n in the subpopulations
4. [**New population**] Create a new population by repeating the following steps:
   Begin
   4.1. [**Selection**] Select parent chromosomes from the subpopulation according to their fitness function
   4.2. [**Crossover**] Cross over the parents to form new offspring with a crossover probability
   4.3. [**Mutation**] Mutate new offspring at each locus with a mutation probability
   End
5. [**Replace**] Use the newly generated loop in an old subpopulation for a further run of the algorithm
6. [**Migration**] Migration of individuals between the subpopulations after following the isolation time
7. [**Test**] If the end condition is satisfied, stop and return the best solution in current population
8. [**Loop**] go to step 2.

End

1.2 Short Overview of MpGA Modifications

Altogether six kinds of MpGA are investigated towards improvement of the algorithms convergence time. All considered MpGA have a similar structure but differ from each other in the sequence of execution of the selection of the main genetic operators, crossover, and mutation. MpGA-SCM (coming from sequence **s**election, **c**rossover,

Table 1.1. Sequence of algorithm steps implementation.

MpGA	*Algorithm steps*
MpGA-SCM	1, 2, 3, *4.1*, *4.2*, *4.3*, 5, 6, 7, 8
MpGA-CMS	1, 2, 3, *4.2*, *4.3*, *4.1*, 5, 6, 7, 8
MpGA-SMC	1, 2, 3, *4.1*, *4.3*, *4.2*, 5, 6, 7, 8
MpGA-MCS	1, 2, 3, *4.3*, *4.2*, *4.1*, 5, 6, 7, 8
MpGA-SC	1, 2, 3, *4.1*, *4.2*, 5, 6, 7, 8
MpGA-CS	1, 2, 3, *4.2*, *4.1*, 5, 6, 7, 8

mutation), MpGA-CMS (**c**rossover, **m**utation, **s**election), MpGA-SMC (**s**election, **m**utation, **c**rossover) and MpGA-MCS (**m**utation, **c**rossover, **s**election) were proposed in [2]. Another two kinds of MpGA avoid a mutation operator. MpGA-SC (**s**election, **c**rossover) is further developed here based on the idea given in [5] for *E. coli* parameter identification. MpGA-CS (**c**rossover, **s**election) is newly developed here, provoked by the promising results obtained when the selection operator is processed after crossover in SGA. Table 1.1 lists the order of the steps for creating a new population for the six kinds of MpGA considered here.

The elaboration of MpGA-CS, newly developed in this investigation, is presented briefly below. The main idea of this modification is that the individuals are reproduced processing only crossover and avoiding mutation. In the beginning, MpGA-CS generates a random population of n chromosomes, i. e., suitable solutions for the problem. In order to prevent the loss of the good solution acquired by crossover, selection has been processed after crossover. Parents' genes combine to form a whole new chromosome during the crossover. After reproduction, the MpGA-CS calculates the objective function for the offspring, and the best-fitted individuals from the offspring are selected to replace the parents, according to their objective value. Then a number of individuals are distributed between the subpopulation (migration). When a certain number of generations is fulfilled, a mean deviation in the population is satisfied, or when a particular point in the search space is encountered, the MpGA-CS is terminated.

There are many operators, functions, parameters, and settings in GA that can be improved and implemented differently in various problems [2, 3, 8]. In this study five of the main genetic algorithms parameters, namely generation gap (GGAP), and rates of crossover (XOVR), mutation (MUTR), insertion (INSR) and migration (MIGR), have been investigated with the values shown in Table 1.2, according to some statements in [6]. A very large generation gap value does not improve the performance of GA, especially regarding how fast the solution will be found. Mutation is randomly applied with low probability, typically in the range 0.01 and 0.1. Crossover rate is the parameter that affects the rate at which the crossover operator is applied. A higher crossover rate introduces new strings more quickly into the population. A low crossover rate may

Table 1.2. Range of investigated genetic algorithm parameters.

GGAP	*XOVR*	*MUTR*	*INSR*	*MIGR*
0.5	0.65	0.02	0.5	0.2
0.67	0.75	0.04	0.6	0.4
0.8	0.85	0.06	0.8	0.6
0.9	0.95	0.08	0.9	0.8
–	–	0.1	1	0.1

cause stagnation due to the lower exploration rate. Insertion rate (INSR) is a general measure of how many of the individuals produced at each population are inserted into the new generation. Migration rate (MIGR) characterizes the number of exchanged individuals.

The focus of the study is the influence of five of the main genetic algorithm parameters, namely generation gap, and rates of crossover, mutation, insertion, and migration, to be investigated with regard to algorithm convergence time. Such genetic algorithm parameters are examined for altogether six kinds of multi-population genetic algorithms. Their performances are demonstrated for *S. cerevisiae* fed-batch cultivation.

1.3 Parameter Identification of *S. cerevisiae* Fed-Batch Cultivation Using Different Kinds of MpGA

Experimental data of *S. cerevisiae* fed-batch cultivation was obtained from the Institute of Technical Chemistry, University of Hannover, Germany [7]. The cultivation of the yeast *S. cerevisiae* is performed in a 1.5 l reactor, using a Schatzmann medium. Glucose in the feeding solution is 50 g/l. The temperature was controlled at 30°C, the pH at 5.7. The stirrer speed was set to 500 rpm. The aeration rate was kept at 300 l/h. Biomass and ethanol were measured offline, while substrate (glucose) and dissolved oxygen were measured online.

The mathematical model of *S. cerevisiae* fed-batch cultivation is commonly described as follows, according to the mass balance [7]:

$$\frac{dX}{dt} = \mu X - \frac{F}{V} X \tag{1.1}$$

$$\frac{dS}{dt} = -q_S X + \frac{F}{V}(S_{in} - S) \tag{1.2}$$

$$\frac{dE}{dt} = q_E X - \frac{F}{V} E \tag{1.3}$$

$$\frac{dO_2}{dt} = -q_{O_2} X + k_L a \left(O_2^* - O_2\right) \tag{1.4}$$

$$\frac{dV}{dt} = F, \tag{1.5}$$

where X – concentration of biomass [g/l], S – concentration of substrate (glucose) [g/l], E – concentration of ethanol [g/l], O_2 – concentration of oxygen [%], O_2^* – dissolved oxygen saturation concentration [%], F – feeding rate [l/h], V – volume of bioreactor [l], k_La – volumetric oxygen transfer coefficient [1/h], S_{in} – initial glucose concentration in the feeding solution [g/l], μ, q_S, q_E, q_{O_2} – specific growth/utilization rates of biomass, substrate, ethanol and dissolved oxygen [1/h].

The fed-batch cultivation of *S. cerevisiae* considered here is characterized with keeping the glucose concentration equal to or below its critical level (S_{crit} = 0.05 g/l), sufficient dissolved oxygen in the broth $O_2 \geq O_{2crit}$ ($O_{2crit} = 18\%$) and availability of ethanol in the broth. This state corresponds to the so-called *mixed oxidative state* (FS II) according to the functional state modeling approach [7]. As presented there, the specific growth rate is generally found to be the sum of two terms, one being the contribution of sugar and the other the contribution of ethanol to yeast growth. Both terms have the structure of the Monod model. The Monod model is also used for the specific ethanol and sugar consumption rates. The dissolved oxygen consumption rate is obtained as the sum of two terms which are directly proportional to the specific glucose rate and specific ethanol production rate, respectively. Hence, specific rates in Equations (1)–(5) are presented as follows:

$$\begin{aligned} \mu &= \mu_{2S}\frac{S}{S+k_S} + \mu_{2E}\frac{E}{E+k_E}, \quad q_S = \frac{\mu_{2S}}{Y_{SX}}\frac{S}{S+k_S} \\ q_E &= -\frac{\mu_{2E}}{Y_{EX}}\frac{E}{E+k_E}, \quad q_{O_2} = q_E Y_{OE} + q_S Y_{OS} \end{aligned} \tag{1.6}$$

where μ_{2S}, μ_{2E} – the maximum growth rates of substrate and ethanol [1/h], k_S, k_E – saturation constants of substrate and ethanol, [g/l], and Y_{ij} – yield coefficients [g/g].

As an optimization criterion, the mean square deviation between the model output and the experimental data obtained during cultivation has been used:

$$J_Y = \sum (Y - Y^*)^2 \to \min, \tag{1.7}$$

where Y is the experimental data, Y^* the model predicted data, and $Y = [X, S, E, O_2]$.

All the computations are performed on a PC Intel Pentium 4 (2.4 GHz) platform running Windows XP. All six kinds of multi-population genetic algorithms have been consequently applied for the purposes of the parameter identification of *S. cerevisiae* fed-batch cultivation. The values of the genetic algorithm parameters except those investigated here and the type of genetic operators for all the kinds of genetic algorithms considered here are tuned according to [2].

The influence of the main genetic algorithm parameters, namely GGAP, XOVR, MUTR, INSR, and MIGR has been examined for all six kinds of MpGA. A comparison in relation to model accuracy and convergence time was performed. Table 1.3 demonstrates the results obtained for the most sensitive parameter GGAP, while the

Table 1.3. Influence of GGAP to model accuracy and convergence time into MpGA with tree genetic operators.

GGAP	*MpGA-SCM*		*MpGA-SMC*		*MpGA-CMS*		*MpGA-MCS*	
	J	*t, s*	*J*	*t, s*	*J*	*t, s*	*J*	*t, s*
0.5	0.0220	100.8910	0.0220	111.7810	0.0221	273.9060	0.0220	307.8440
0.67	0.0221	112.1720	0.0220	141.0940	0.0221	325.5780	0.0220	332.0620
0.8	0.0221	155.4680	0.0220	178.9680	0.0221	321.0160	0.0221	373.1560
0.9	0.0220	170.2660	0.0220	340.6720	0.0221	343.6870	0.0221	349.7500

Table 1.4. Influence of GGAP to model accuracy and convergence time into MpGA with two genetic operators.

GGAP	*MpGA-CS*		*MpGA-SC*	
	J	*t, s*	*J*	*t, s*
0.5	0.0223	267.9220	0.0222	111.5310
0.67	0.0222	331.9690	0.0223	119.7340
0.8	0.0223	333.6250	0.0221	153.3900
0.9	0.0221	357.0160	0.0220	168.2190

results for XOVR, MUTR, INSR, and MIGR are taken into account in the analysis. Because of the stochastic nature of GA, several runs were performed in order to achieve representative results. The best ones obtained are presented here. As shown in Tables 1.3 and 1.4, the optimization criterion values obtained with six kinds of multi-population genetic algorithms are very similar, varying between 0.0220 and 0.0224, which means less then 2% divergence. This fact is very promising due to the fact that there is no loss of accuracy. It can be noted that the results obtained with MpGA-SCM are very similar to those obtained with MpGA-SMC. Also, when MpGA-CMS is applied, the results are close to those obtained with MpGA-MCS, but the convergence time is much greater than the first group. It is also demonstrated, that there is no loss of model accuracy when the operator mutation is performed before crossover. It can be summarized that proceeding selection operator before crossover and mutation (in whatever order) needs much less computational time. This fact is valid for all investigated genetic algorithm parameters, i. e., GGAP, XOVR, MUTR, MIGR, and INSR. Two algorithms without mutation execution, MpGA-SC and MpGA-CS, can also be grouped together. In cases when algorithms are only implemented with two operators (selection and crossover), the accuracy logically decreases.

It should be noted that the GGAP is the most sensitive of the five investigated parameters with regard to convergence time. Up to 40% (in case of MpGA-SCM, which is distinguished also as being the fastest algorithm) can be saved using GGAP = 0.5 instead of 0.9 without loss of accuracy. But if one compares MpGA-SCM where GGAP = 0.5 (the fastest one) to MpGA-MCS where GGAP = 0.8 (the slowest one), it is almost 3.7 times faster with even higher model accuracy. Exploring different values of crossover rate, no such time saving is realized, but it should be pointed out that values of 0.85 for XOVR can be assumed as more appropriate for all considered algorithms, since at this value two of the algorithms are the fastest while another four are the second fastest. Examining XOVR, the fastest algorithm is MpGA-SMC at XOVR = 0.85, it achieves the solution even faster than the standard MpGA-SMC. If one compare this fastest algortihm to the slowest one here, MpGA-CMS at XOVR = 0.75, it is almost three times faster. Exploring MUTR, three of four considered algorithms are the fastest ones at MUTR = 0.02 and just one is faster at MUTR = 0.04, but the time is very close to that at MUTR = 0.02. This is why MUTR = 0.02 can be assumed, saving up to 46% in MpGA-SMC without loss of accuracy. In INSR and MIGR no tendency of influence can be seen. The comparison presented here shows that the implementation of the operators in a sequence selection and mutation followed by crossover or in a reverse order for the last two operators is the optimal one with regard to convergence time with guarantied high accuracy of the decision. For three of the considered genetic parameters, MpGA-SCM is the fastest, while for another two parameters it is the second fastest. Because of the similarity of the results obtained with all six kinds of algorithms (as mentioned, with 2% divergence in model accuracy), the results obtained by the MpGA-CS developed here, even though it is not the fastest one, are presented.

The genetic parameter values chosen based on the above analysis are as follows: GGAP = 0.5, XOVR = 0.85, MUTR = 0.02, INSR = 0.9, and MIGR = 0.1.

As a result of parameter identification, the values of the model parameters are respectively: $\mu_{2S} = 0.98$ [1/h], $\mu_{2E} = 0.13$ [1/h], $k_S = 0.13$ [g/l], $k_E = 0.84$ [g/l]; $Y_{SX} = 0.42$ [g/g], $Y_{EX} = 1.67$ [g/g], $k_La = 96.23$ [1/h];,$Y_{OS} = 766.79$ [g/g], and $Y_{OE} = 125.52$ [g/g], the CPU time being 288.6720 s and J = 0.0221.

Figure 1.1 presents results from experimental data and model prediction respectively for biomass, ethanol, substrate, and dissolved oxygen.

The results presented from MpGA-CS application for parameter identification of *S. cerevisiae* fed-batch cultivation show the effectiveness of GA for solving complex nonlinear problems.

1.4 Analysis and Conclusions

In this investigation, altogether six kinds of MpGA have been examined. Four different modifications of MpGA are with exchanged operators' sequence of selection, crossover, and mutation operators. Two MpGA modifications are without performing the mutation operator. The influence of some of GA parameters, namely generation gap, crossover, mutation, insertion, and migration rates, has been examined for all six kinds of genetic algorithms with the aim of improve the convergence time. Among the five investigated parameters, the generation gap is the most sensitive one with regard to convergence time. As "favorites" among the algoritms considered here with

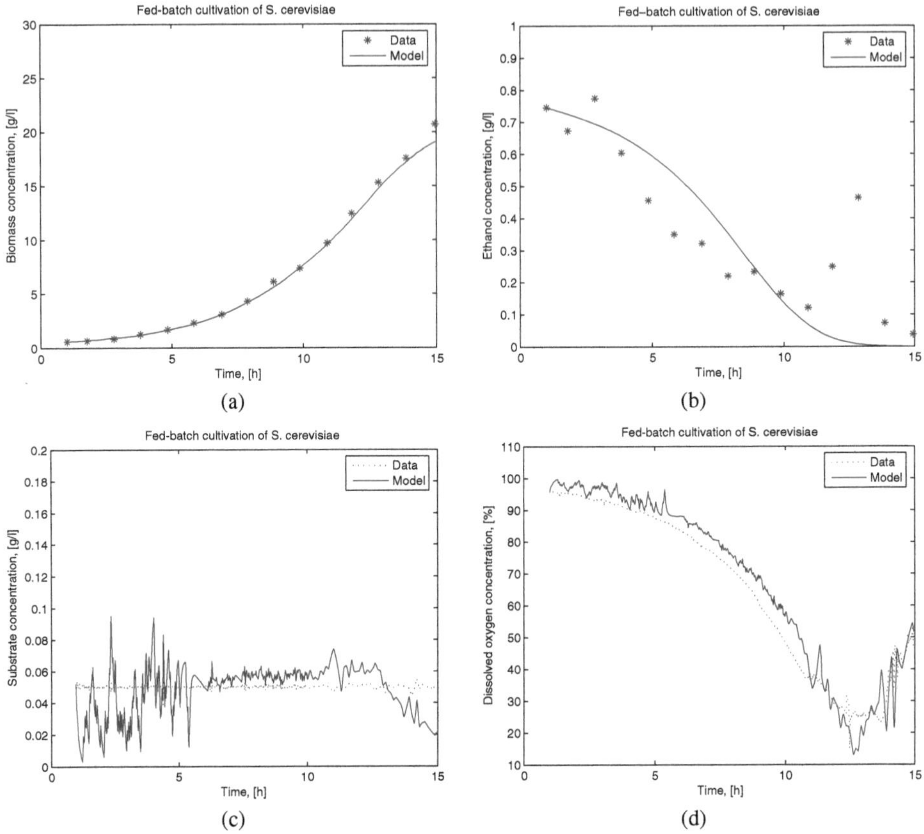

Figure 1.1. Model prediction compared to experimental data.

two and three genetic operators performance, MpGA-SCM has been distinguished as the fastest one. Up to almost 40% of the calculation time can be saved in the case of MpGA-SCM application using GGAP = 0.5 instead of 0.9 without loss of model accuracy. Exploring the different values of crossover, mutation, insertion, and migration rates yielded no such time saving. But when one compares different modifications of MpGA with different values of GA parameters, up to 4.7 times of calculation time can be saved. Overall, all modification of MpGA employing different values of GA parameters show the effectiveness of genetic algorithms for solving complex nonlinear problems.

Acknowledgments. This work is partially supported by the European Social Fund and Bulgarian Ministry of Education, Youth and Science, grant BG051PO001-3.3.04/40 and National Science Fund of Bulgaria, grant DID 02-29.

References

[1] J. Adeyemo and A. Enitian, Optimization of fermentation processes using evolutionary algorithms – a review, *Scientific Research and Essays* **6** (2011), 1464–1472.

[2] M. Angelova, S. Tzonkov, and T. Pencheva, Genetic algorithms based parameter identification of yeast fed-batch cultivation, in: *Proc. of Numerical Methods and Applications*, Lect. Notes Comp. Sci. 6046, pp. 224–231 2011.

[3] D. Goldberg, *Genetic algorithms in search, optimization and machine learning*, Addison-Wiley Publishing Company, Boston, 1989.

[4] K. Jones, Comparison of genetic algorithms and particle swarm optimization for fermentation feed profile determination, in: *Proc. of CompSysTech'2006*, Veliko Tarnovo, Bulgaria, 15–16 June, 2006, IIIB.8-1-IIIB.8-7.

[5] K. Kosev and O. Roeva, *Modified multi-population genetic algorithm with improved convergence*, 8th National Young Scientific-Practical Session, Sofia, Bulgaria, May 10–11, 2010, pp. 7–12 (in Bulgarian).

[6] M. Obittko, *Genetic algorithm*, (2005), available online at http://cs.felk.cvut.cz/~xobitko/ga/main.html (last accessed September 21, 2012).

[7] T. Pencheva, O. Roeva, and I. Hristozov, in: S. Tzonkov and B. Hitzmann (eds.), *Functional State Approach to Fermentation Processes Modelling*, Prof. Marin Drinov Academic Publishing House, Sofia, 2006.

[8] O. Roeva, *Improvement of genetic algorithm performance for identification of cultivation process models*, Advanced Topics on Evolutionary Computing, Book Series: Artificial Intelligence Series-WSEAS, pp. 34–39, 2008.

[9] P. Vasant and N. Barsoum, Hybrid genetic algorithms and line search method for industrial production planning with non-linear fitness function, *Engineering Applications of Artificial Intelligence* **22** (2009), 767–777.

Author information

Maria Angelova, Institute of Biophysics and Biomedical Engineering, Bulgarian Academy of Sciences, Sofia, Bulgaria.
Email: `maria.angelova@clbme.bas.bg`

Tania Pencheva, Institute of Biophysics and Biomedical Engineering, Bulgarian Academy of Sciences, Sofia, Bulgaria.
Email: `tania.pencheva@clbme.bas.bg`

Monte Carlo Methods and Applications, 11–20

Chapter 2

Parallelization and Optimization of 4D Binary Mixture Monte Carlo Simulations Using Open MPI and CUDA

Sergey Artemchuk and Paula A. Whitlock

Abstract. A Metropolis Monte Carlo code that studies the properties of a binary mixture of four dimensional hyperspheres is modified to run on a network of hosts with two Nvidia GPU attached to each host. Significant decreases in execution time are achieved as a result of the code modifications, which include the use of OpenMPI and the CUDA extensions to the C computer language.

Keywords. Monte Carlo, Hard Hyperspheres, Four Dimensions, MPI, GPU, CUDA.

Mathematics Subject Classification 2010. 68M14, 68W15, 80M31.

2.1 Introduction

Excluded volume interactions between granular objects are a determining factor in the structure and properties of most fluid and solid systems. These interactions can be modeled by considering systems composed of hard spheres. As the dimensionality of the system increases to four or more dimensions, the number of possible lattices increases and the thermodynamic properties are more complex. However, at very high spatial dimension [1], systems parallel the effects of low density regimes and the thermodynamic properties becomes simpler. Thus systems in dimensions greater than three are of intrinsic interest. The properties of pure fluids and solids of hard hyperspheres in four dimensions have been well studied in the past few years [2–5], but the properties of mixtures of differing size ratios of the hyperspheres have been less often investigated [6]. A research project was initiated to study two component mixtures of hyperspheres in four dimensions using a Metropolis Monte Carlo (MC) method. The simulations are straightforward, but to minimize size effects all the random walks involve a large number of hyperspheres. Running multiple random walks in a multithreaded calculation on quadcore AMD machines has proven to be slow, often taking several weeks of almost continuous computation to reach the equilibrium distribution [7]. This bottleneck provided the impetus to rewrite the multithreaded code to evaluate the advantage of using multiple Nvidia GPU Fermi's available on a network of CPU hosts [8].

This chapter describes the strategies used to first distribute the four dimensional code over a network using OpenMPI and then perform most of the Metropolis algorithm on a GPU attached to each host machine programmed with the CUDA extensions to the C language. Section 2.2 gives a brief description of the MC method. In Section 2.3, the domain decomposition using OpenMPI is discussed. Section 2.4 explains the computations that are sent to the GPUs on each host. Section 2.5 discusses how pseudorandom numbers were assigned on the GPUs and finally, in Section 2.5 results for the model system are given.

2.2 The Metropolis Monte Carlo Method

The Metropolis [9] method was developed as a sampling method for the Monte Carlo evaluation of integrals involving complex physical systems. It guarantees the asymptotically correct sampling of a probability distribution function by performing random walks in the many dimensional configuration space of the problem. Here, the behavior of the d-dimensional hyperspheres is governed by the Boltzmann distribution function, $f(\mathbf{R})$:

$$f(\mathbf{R}) = \frac{\exp[-\sum \phi(r_{ij})/k_b T]}{\int \exp[-\sum \phi(r_{ij})/k_b T] d\mathbf{R}}, \tag{2.1}$$

where $\mathbf{R}$ is the d-dimensional vector composed of the d*M coordinates of the centers of mass of the M hyperspheres, k_b is Boltzmann's constant and T is the absolute temperature of the system. The pair potential $\phi(r_{ij})$ represents the interaction between two hyperspheres, and r_{ij} is the hypersphere separation.

The hyperspheres are initially placed in a simulation box on a lattice with periodic boundary conditions. The side lengths of this box are determined by the number of hyperspheres, N, per side, and the number density, ρ, the number of particles per unit volume.

Random walks are generated by proposing a move from the current position of a hypersphere, $\mathbf{X}$, to a new position, $\mathbf{X}'$. The new position is chosen from a probability distribution function, $H(\mathbf{X}' \mid \mathbf{X})$. In the present calculation, the new position is randomly chosen from a hyperbox surrounding the current position of the center of mass of the hypersphere. The proposed new position is then accepted or rejected based upon the probability $p(\mathbf{X}' \mid \mathbf{X})$ related to $f(\mathbf{R})$ [10]. As the random walk proceeds, a recursive relationship develops between the phenomenological distribution functions, $f_n(\mathbf{R})$, represented by each step of the random walk. As long as the system is ergodic and obeys detailed balance [9–11], $f_n(\mathbf{R}) \to f(\mathbf{R})$ is guaranteed to be true as n, the number of steps, becomes large.

Since the successive positions of the hyperspheres are not independent, it takes many passes, one attempted move by each of the hypersphere, to converge from the initial state to an equilibrated one sampled from $f(\mathbf{R})$. Once the asymptotic distribution function is reached, there is still serial correlation between each step in the random walk. Therefore, the random walks are divided into blocks and the block averages are used to calculate any quantities of interest.

2.3 Decomposition into Subdomains and the Virtual Topology Using OpenMPI

Dividing a problem into subdomains is a well-known method for distributing the simulations of large physical systems over a network of computers. The use of subdomains is also a technique to decrease the number of pairwise comparisons that need to be performed, although it comes at a cost of increased communication between the CPUs on the network. Here, the four-dimensional space was divided into subdomains, and each subdomain was distributed to a different CPU host. The division was accomplished by representing the four-dimensional space by a three-dimensional grid where the fourth dimension coordinate, w, will always stay with a particular CPU as long as its x, y, z coordinates stay within the subdomain's boundaries. However, once the x-, y- or z-coordinate moves outside the boundary of the subdomain, the hypersphere's coordinate data will be transfered from the current host to another one that contains the new subdomain.

The Metropolis MC calculation involves a pseudodynamics which proposes a possible move of a hypersphere. The move is accepted if the new position does not overlap with another hypersphere. Otherwise the move is rejected. Within a subdomain, overlap is straightforward to determine, although computationally intensive. However, to verify whether any overlap has occurred between hyperspheres on different hosts, coordinate data must be transferred between the hosts. To minimize communication, "ghost boundaries" are established that are interior to the actual subdomain boundaries. If a hypersphere is located between the ghost boundary and the subdomain boundary, its coordinates will be temporarily transferred to the neighboring subdomain on another host to check for overlap. How much data needs to be transferred in the worst case when using the ghost boundaries? In three-dimensional simulations with subdomains, for each side of the cube, N^2 spheres may need to be transferred. However because this is a four-dimensional system, N^3 per side or $6 * N^3$ hyperspheres may need to be transferred to other CPUs and the same number may be received back. Since the majority of the numerical computations are performed on the Nvidia fermi GPU, each time there is a transfer of hyperspheres from one host CPU to another, the data must be made available to the GPU. This entails yet another step in the data transfer.

2.4 Management of Hypersphere Coordinate Migration Between Domains

Because the system is dynamic and involves constant moves of the hyperspheres between the host CPU and GPU, the calculation needed to be implemented to fully exploit the computational power of the GPU. The data structures used on the GPU have been designed to allow management of the calculation by multiple threads. Four main buffers for the x, y, z, and w coordinates are maintained and each buffer is split into eight sectors. The first sector of each buffer is called DATA. It is a place where the coordinates of the hyperspheres that currently belong to the subdomain are stored. The next sector is RESERVED SPACE for the coordinates of newly received hyperspheres from another subdomain. Finally, there are six sectors that are used for the ghost boundaries. In addition to the x, y, z, and w buffers, there are six additional buffers that represent the occupancy of the hyperspace by hyperspheres. The management of these data structures is helped by the use of a stream compaction algorithm with a prefix summation [12]. In the compaction algorithm, the hypersphere coordinate arrays and the occupancy arrays are of particular importance. In the occupancy arrays, a '1' indicates that the hypersphere will leave the current subdomain, and a '0' means it will stay. By applying the prefix sum algorithm on the occupancy array, the new indexes for all hyperspheres are determined, and they may be moved to the right location. This is an effective way of managing the coordinates, because once a hypersphere is marked as staying by '0', it can be removed from the GPU data partition that represents hyperspheres which will be moved.

The prefix sum algorithm was implemented to handle large array sizes [13] through a fragmented prefix sum approach. The array was divided into `Array_size` / `(block_size*2)` partitions, where `Array_size` is the number of elements in the original array and `block_size` is the number of elements to be processed by a block of thread processors. The blocks of thread processors on the GPU perform the prefix sum on their own data partitions. Once each block has completed its sum, the result of their calculations, a single number, will be written into an auxiliary array. Another prefix sum will be performed on this auxiliary array, and thus the final sum will be obtained from all the blocks. The segmented data partitions are updated with the new values by adding corresponding sums to each element in the corresponding partition. On large arrays this algorithm on the GPU is about 3–5 times faster then on the host CPU.

If a hypersphere has crossed a ghost boundary after a move, a '1' is written into the occupancy buffer with the corresponding index. For example, hypersphere A with index 10 has crossed the ghost boundary labelled TOP. Then '1' will be stored in the occupancy array TOP at index 10. After all the hyperspheres are moved and the occupancy buffers are populated, a prefix sum is performed and the positions within the occupancy buffers are calculated. If a data element has '1' in one of the occupancy buffers, it is copied to the corresponding boundary buffer with a precalculated index from the prefix sum. When the boundary buffers are ready, the host will take over and copy them into its memory.

2.4.1 Communication between the CPU and the GPU

Each time data is shared between different hosts, the data needs to be transferred to the associated GPU. For example, data transfer between the GPU and the host must occur before the host performs a `MPI_Send`.

When a random move is attempted, the new coordinates are compared to the ghost boundaries. If any hypersphere crosses into the ghost zone, it is placed into a corresponding buffer for transfer to the host. The host will copy the coordinates into corresponding host memory locations and perform the `MPI_Send` to its neighbors.

Similarly, once a `MPI_Recv` receives hypersphere coordinates from a neighboring host, the corresponding data will be copied into the GPU device memory. The GPU then checks for overlap with any hyperspheres within its subdomain. Depending on the magnitude of an accepted move, the hypersphere may be permanently transferred to the new subdomain.

2.5 Pseudorandom Number Generation

A good pseudorandom number generator is extremely important for the simulation. Each host/GPU pair needs a sequence of pseudorandom numbers independent of the other host/GPU pairs. The Nvidia SDK [14] implements the Mersenne Twister algorithm for use on a single GPU. However, independent versions of the parameters are needed for the distributed calculation. An application library, the Dynamic Creator of Mersenne Twisters dcmt0.6.1 [15], is available to generate multiple instances of different pseudorandom number sequences and has been used here. To obtain independent sequences, the parameters are pregenerated using the Dynamic Creator application in a separate code and stored in a file. The host identified as process 0 in the MPI communicator opens the file, reads the contents and uses `MPI_Send` to distribute the parameters to each host. Once the contents of the file are distributed to all the processes, each host passes the received parameters and configuration data, to its GPU. In the GPU, the multiple blocks execute the MersenneTwisterGPU kernel that generates a set of pseudorandom numbers of a specified size using multiple threads. For future use by the other kernels in the program, the MersenneTwisterGPU kernel returns a pointer to global memory that holds a set of pseudorandom numbers.

2.6 Results of Running the Modified Code

The rewritten code was checked after each change to validate that it still correctly performed the Metropolis MC calculation. One goal in rewriting the code was to achieve

Table 2.1. Timings of simulations with differing numbers of threads for 800 passes on one host/GPU pair.

Number of hyperspheres		256	512	1024
computer code	# of threads	time(s)	time(s)	time(s)
C program on host	1	6.5	26.73	109.815
Cuda code on GPU	1	75.56	296.812	1196.316
Cuda code on GPU	32	5.37	17.375	65.882
" " " "	64	3.09	9.537	34.112
" " " "	128	1.96	5.780	18.739
" " " "	256	1.32	3.773	12.415
" " " "	512	–	2.797	8.905

a speedup over the serial code running on a single CPU host. The simulations were run on a network of 48 8-core SGIX348 hosts with two Nvidia Fermi GPUs per host. There are a total of 96 GPUs available in the network. The hosts on the network are connected by a 40 Gbit/s QDR infiniband. In Table 2.1, the timings for running the simulation for 800 passes with two codes, the serial code on a host vs. parallel code using a single GPU, are compared for a small number of hyperspheres. The power of the multithreaded architecture of the Nvidia Fermi is most clearly utilized when the thread processors perform many repeated operations on large data sets. This fact is demonstrated in Table 2.1 as the number of hyperspheres in increased. In all cases, the parallel code on the GPU executes much slower than the serial code on the host when only one thread is used. This behavior is not unexpected, since the transferring of data between the host and the GPU adds a significant communication cost and no benefit from simultaneous calculations by many threads. However, as soon as multiple threads are employed in groups of 32 (called a warp), speedups are achieved. With 256 hyperspheres, and 256 threads, the calculation is almost five times faster than the serial calculation for the same number of passes. Increasing the number of hyperspheres to 512 and the number of threads on the GPU to 512 yields a calculation nine times faster than on the host. And finally, increasing the number of hyperspheres to 1024 and still using 512 threads yields a calculation more than 12 times faster.

A speedup of 12 times faster is good, but the possibility of greater speedups are possible by distributing the calculation over a network. In Table 2.2, timings are given for a variety of sized systems distributed over a network of 8 CPU/GPU pairs. Each subdomain has two (2) hyperspheres on each side. Each calculation is run for 800 passes. Table 2.2 shows the communication time as well as the total computation time. The communication time is a large percentage of the total time, but it is acceptable, since the speed-up is spectacular. For comparison, a calculation with 10^4 hyperspheres run on an AMD quadcore machine for 800 passes took 16,200 seconds. The calculation using eight CPU/GPU pairs was more than 186 times faster!

Table 2.2. Timings of the simulation with subdomains spread across a network of 8 CPU/GPU pairs.

System size	*Number of subdomains per side*	*IO time (s)*	*Total time (s)*
4X4X4X4	2	5.17	8.59
6X6X6X6	3	14.83	20.41
8X8X8X8	4	33.87	45.21
10X10X10X10	5	65.41	86.93
12X12X12X12	6	112.13	166.58
14X14X14X14	7	179.15	274.77
16X16X16X16	8	268.81	440.75
18X18X18X18	9	384.14	644.35

Table 2.3. Timings of simulations with subdomains spread across a network of 27 CPU/GPU pairs.

System size	*IO time (s)*	*Total time (s)*
6X6X6X6	12.32	15.54
9X9X9X9	31.52	38.42
12X12X12X12	68.41	81.67
15X15X15X15	132.83	168.43
18X18X18X18	227.49	327.84
21X21X21X21	365.19	548.77
24X24X24X24	538.32	873.22
27X27X27X27	669.28	1282.76
30X30X30X30	1058.93	1739.54

With the parallelization of a computer code, there is always the question of whether the code will scale well as both the number of nodes in the distributed computation and the size of the problem increases. Experiments were performed in which both the number of CPU/GPU hosts were increased and the number of hyperspheres were increased. Table 2.3 displays the timing results of distributing the four-dimensional calculation over a network of 27 CPU/GPU pairs. Again, each simulation is run for 800 passes.

There is clearly an advantage to using a larger number of nodes even for smaller systems of 20,736 hyperspheres (12 hyperspheres per side of the simulation box). Using 27 nodes, the total execution time was 81.67 s compared to a time of 166.58 s when 8 nodes were used. The network that is used allows for even a larger number of

Table 2.4. Timings of simulations with subdomains spread across a network of 64 CPU/GPU pairs.

System size	*IO time (s)*	*Total time (s)*
8X8X8X8	13.39	20.39
12X12X12X12	32.71	41.66
16X16X16X16	72.32	90.99
20X20X20X20	134.33	175.08
24X24X24X24	232.15	339.28
28X28X28X28	365.12	553.00
32X32X32X32	540.78	881.54
36X36X36X36	775.98	1290.05
40X40X40X40	1065	1749.18

CPU/GPU pairs and the results for using 64 nodes is shown in Table 2.4. It is immediately apparent that the system with 20,736 hyperspheres again has a shorter execution time, 41.66 s, than either of the previous calculations with 8 or 27 nodes.

The timings given in Tables 2.2–2.4 are graphically represented in Figure 2.1. The results from Table 2.2 are displayed using solid squares; solid circles represent the results from Table 2.3; and solid diamonds represent the outcomes from Table 2.4. The lines are present to guide the eye. When system sizes can be compared, more nodes give faster times even though the communication time dominates each calculation.

2.7 Conclusions

The goal of this research was to effectively use GPUs to parallelize a Metropolis Monte Carlo code. While the MC algorithm was straightforward to understand, the fact that every hypersphere in the four dimensional system could interact with every other hypersphere made the computational complexity on the order of N^2. Moving most of the computation to the GPU enabled the testing for overlap when a hypersphere was moved to be done in parallel by multiple threads. Just this change alone to the code achieved a speedup of a factor of twelve. Then, for large N systems, partitioning the domain into subdomains and distributing these on a fast network lead to even greater speedups, e. g., a minimum of a factor of 186.

Future improvements to the code will incorporate the use of streams to increase the efficiency of data transfer between the host CPU and the GPU. Additional optimizations can occur by using a more efficient stream compaction algorithm [16]. Finally, the slowest part of the simulation is still the determination of possible overlap when a

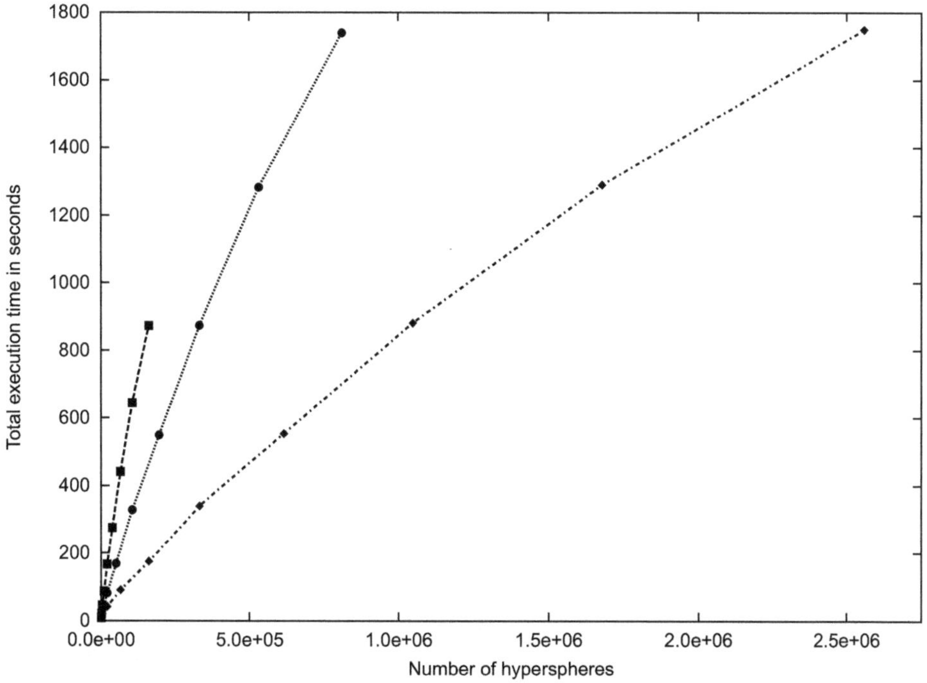

Figure 2.1. The timings for simulations as a function of the number of hyperspheres. The solid squares represent the use of 8 CPU/GPU pairs, the solid circles are for 27 CPU/GPU pairs, and the solid diamonds are for 64 CPU/GPU pairs.

hypersphere is moved. Incorporating warp vote functions is one method that can make overlap detection more efficient.

Acknowledgments. This research was supported in part under National Science Foundation Grants CNS-0958379 and CNS-0855217 and the City University of New York High Performance Computing Center. One of us, P.A.W., thanks Professor Marvin Bishop, Manhattan College for sharing his serial Metropolis Monte Carlo code with us and for many useful conversations.

References

[1] H. L. Frisch and J. K. Percus, High dimensionality as an organizing device for classical fluids, *Phys. Rev. E* **60** (1999), 2942–2948.

[2] M. Skoge, A. Donev, F. H. Stillinger, and S. Torquato, Packing Hyperspheres in High-Dimensional Euclidean Spaces, *Phys. Rev. E* **74** (2006), 041127.

[3] M. Adda-Bedia, E. Katzav, and D. Vella, Solution of the Percus-Yevick equation for hard hyperspheres in even dimensions, *J. Chem. Phys.* **129** (2008) 14450.

[4] J. A. van Meel, D. Frenkel, and P. Charbonneau, Geometrical Frustration: a Study of 4d Hard Spheres, *Phys. Rev. E* **79** (2009), 030201.

[5] L. Lue, M. Bishop, and P. A. Whitlock, The Fluid to Solid Phase Transition of Hard Hyperspheres in Four and Five Dimensions, *J. Chem. Phys.* **132** (2010), 104509.

[6] A. Santos, S. B. Yuste, and M. Lopez de Haro, Virial coefficients and equations of state for mixtures of hard discs, hard spheres and hard hyperspheres, *Mole. Phys.* **99** (2001), 1959–1972.

[7] M. Bishop and P. A. Whitlock, Monte Carlo Study of Four Dimensional Hard Hypersphere Mixtures, *J. Chem. Phys.* **36** (2012), 014506.

[8] M. Garland, D. B. Kirk, Understanding Thoughput-oriented Architectures. *Comm. of the ACM* **53** (2010), 58–66.

[9] N. Metropolis, A. W. Rosenbluth, M. N. Rosenbluth, A. H. Teller, and E. Teller, Equations of state calculations by fast computing machines, *J. Chem. Phys.* **21** (1953), 1087–1091.

[10] M. H. Kalos and P. A. Whitlock, *Monte Carlo Methods*, 2nd ed., Wiley-VCH Verlag GmbH & Co., Weinheim, 2008.

[11] W. W. Wood, Monte Carlo studies of simple liquid models, in: H. N. V. Temperley, J.S. Rowlinson and G. S. Rushbrooke (eds.), *The Physics of Simple Liquids*, Chap. 5, North-Holland, Amsterdam, 1968.

[12] M. Billeter, O. Olsson, and U. Assarsson, Efficient stream compaction on wide SIMD many-core architectures, in: *Proceeding HPG '09 Proceedings of the Conference on High Performance Graphics 2009*, ACM, New York, NY, 2009.

[13] M. Harris, S. Sengupta, and J. D. Owens, Parallel Prefix Sum (Scan) with CUDA, in: H. Nuguyen (ed.), *GPU Gems 3*, Chap. 39, Addison-Wesley Professional, Boston, 2007.

[14] Nvidia SDK, Mersenne Twister, available online at http://developer.nvidia.com/cuda/cuda-libraries-sdk-codesamples#MersenneTwisterGP11213 (last accessed September 14, 2012).

[15] M. Matsumoto and T. Nishimura, Dynamic Creation of Pseudorandom Number Generators, in: *Proc. Monte Carlo and Quasi-Monte Carlo Methods 1998*, pp. 56–69, Springer, 2000.

[16] cudapp, available online at http://code.google.com/p/cudpp/ (last accessed September 14, 2012).

Author information

Sergey Artemchuk, Department of Computer and Information Sciences, Brooklyn College, Brooklyn, NY, USA.
Email: `sartemch@gmail.com`

Paula A. Whitlock, Department of Computer and Information Sciences, Brooklyn College, Brooklyn, NY, USA.
Email: `whitlock@brooklyn.cuny.edu`

Chapter 3

Efficient Implementation of the Heston Model Using GPGPU

Emanouil Atanassov, Dimitar Dimitrov, and Sofiya Ivanovska

Abstract. The Heston stochastic volatility model is widely used for the modeling of option prices in financial markets. By adding a jump process to the model, one can account for large spikes in volatility and achieve a better fit of the implied volatility surface. When the parameters of the model have been calibrated to the observed market prices, the model can be used to compute prices of exotic options by Monte Carlo or quasi-Monte Carlo simulations.

In our work we concentrate on the efficient implementation of the schemes of Kahl-Jäckel and Andersen while using the scrambled Sobol' and Halton sequences. The codes were developed using CUDA for NVIDIA GPUs. We apply our methods to the problem of computing the Sobol' sensitivity indices of the option prices as a function of the parameters of the Heston model and present numerical and timing results.

Keywords. Option Pricing, Low-Discrepancy Sequences, GPGPU Computing.

Mathematics Subject Classification 2010. 65C05.

3.1 Introduction

The problem of the numerical estimation of prices of financial assets is one of the most important topics in financial mathematics. The pricing of financial instruments, whose price is dependent on the evolution of the price of other instruments, called "underlying", relies on the precise simulation of the stochastic processes that model the underlying. Such complex instruments, called "derivatives", are traded widely on regulated exchanges or over-the-counter. One of the most popular type of derivatives are the financial options [8, 9]. The financial option is a contract that gives the owner the right but not the obligation to buy or sell an asset at a set price on or before a given date. In this paper we limit our scope by considering only options with a single underlying asset S. At a specified time T the buyer receives payout, which is a function F of the evolution of the price S during time interval $[0, T]$. For a European call option the payout has the form $F = \max(0, S(T) - K)$, where K is the strike price and T is the time of expiry. American call options allow early exercise, but for the purpose of price estimation this is important only if dividend payment is possible within the time period. The prices of more complex (also called "exotic") options

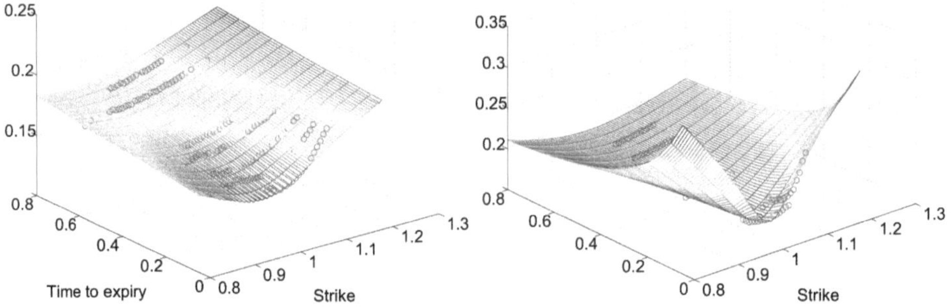

Figure 3.1. Implied volatility as a function of normalized strike price and time to expiration (in years), obtained in two different days using the classic Heston model (on the left) and the Heston model with jumps (on the right).

depend also on the intermediate prices $S_t, t \in [0, T]$. In this chapter we consider as examples several of the most popular exotic options, although our approach is general and can be applied to all path-dependent options. An Asian call option has payoff of the form $F = \max(0, avg(S(t)) - K)$. A barrier knock-out option has payoff of the form $F = \max(0, S(T)) - K)$ but only if the price never exceeds quantity L, otherwise $F = 0$. The mathematical theory of estimation of option prices is based on the idea that there is unique risk neutral measure Q, under which the discounted prices are Martingales: $\exp^{-rt} S_t = E_Q[\exp^{-rT} S_T|\mathcal{F}_t]$ where r is the interest rate. Under these assumptions, today's fair value of the option is $\exp(-rT)E_Q F$. Numerical estimation of this price requires efficient simulation of the evolution of the price S with a sufficiently small time step. Under the Black–Scholes model the price of the asset is lognormally distributed: $S_t = S_0 \exp((r - \frac{1}{2}\sigma^2)t + \sigma W_t)$ where S_0 is the initial stock price, W_t is the Brownian motion, r is the risk-free interest rate, and σ^2 is the "volatility". For each maturity and strike price one can obtain the value of volatility, also called "implied volatility", for which the Black–Scholes formula gives the observed price. In this way for each time-to-expiration and strike price, we obtain data points for the "implied volatility surface". The models for the evolution of the price of the underlying attempt to match the observed "implied volatility surface". In Figure 3.1 one can see two examples of implied volatility surfaces obtained for two different trading days and using different models. The data points are visible as red circles.

The Black–Scholes formula is simple and easy to implement but is not sufficient for adequate modeling of the evolution of asset prices. More complex models have been developed to overcome its limitations. One of these models is the Heston model [10], which considers a stochastic process for the evolution of volatility:

$$dX(t)/X(t) = rdt + \sqrt{V(t)}dW_X(t)$$

$$dV(t) = \kappa(\theta - V(t))dt + \varepsilon\sqrt{V(t)}dW_V(t),$$

where $\kappa, \theta, \varepsilon$ are constants, $X(t)$ the asset price process, $V(t)$ instantaneous variance, and W_X, W_Y Brownian motions. The initial conditions are $X(0) = X_0$ and $V(0) = V_0$. We assume that $\langle dW_X(t), dW_Y(t)\rangle = \rho dt$, where ρ is a correlation parameter. Under the Heston model the prices of European call options can be obtained efficiently, e. g., by using a method based on Fourier transform [7]. Thus model calibration, using optimization procedures like adaptive simulated annealing, is fast and applicable in practice. The Heston model still has some problems to appropriately match the observed form of the implied volatility surface. For long times to expiration the fit is acceptable, but for short times to expiration it is considered insufficient. Various approaches for further improvement of the Heston model have been proposed. One widely-used method is to extend the model by adding the possibility of jumps in the underlying. In this way the first equation becomes

$$dX(t)/X(t) = (r - \lambda\hat{\mu_J})dt + \sqrt{V(t)}dW_X(t) + (J-1)dq.$$

The term $(J-1)dq$ corresponds to a Poisson process describing jumps in the price of the underlying with jump intensity λdt and mathematical expectation $\hat{\mu_j}$. The distribution of jumps can be lognormal or double exponential. This extension of the Heston model adds more parameters, but the model calibration is still numerically efficient, using a similar FFT-based method [8]. After obtaining the model parameters to fit the observed European options' prices, one can proceed to compute the prices of path-dependent exotic options. The most widely applicable method for computing these prices is the Monte Carlo method, where one discretizes along the time and simulates the evolution of the price of the underlying. Many schemes have been developed which attempt to obtain a good balance between complexity and accuracy. In this chapter we consider two of them. The scheme of Kahl–Jäckel–Milstein is an extension of the classic scheme of Milstein, and provides alternative discretization for the price of the underlying [12]. The scheme of Andersen [1] sacrifices the full unbiasedness achieved under the scheme of Broadie and Kaya [6] to attain much faster execution with similar accuracy. The addition of jumps in the underlying is also discussed by Andersen, where he proposes a Martingale-corrected scheme, which he denotes as QE, for the discretized stock price process. In the next section we describe our GPGPU-based algorithm and computer codes implementing these two schemes. Numerical results achieved using these codes are shown in Section 3.3, where we also discuss their application for computing the Sobol‘ sensitivity coefficients for the Heston model with jumps.

3.2 Our GPGPU-Based Algorithm for Option Pricing

The General Purpose GPU computing uses powerful graphics cards to achieve power- and cost-efficient computations and is becoming an ever more popular tool for solving financial mathematics problems. State of the art graphics cards have a large number of transistors and provide hundreds and even thousands of computational cores, and are especially efficient for Monte Carlo simulations because there is a large degree of independence between separate numerical trajectories, and the amount of synchronization to be performed is relatively small. Quasi-Monte Carlo algorithms benefit from GPGPU computing in a similar fashion if efficient algorithms for generation of the low-discrepancy sequences are available. Two approaches are widely used for parallelizing quasi-Monte Carlo methods: "blocking" and "leap-frogging".

In our work we used graphics cards designed by NVIDIA which support the CUDA parallel computing architecture [18]. To achieve high efficiency on CUDA-supporting devices, one has to split the bulk of the computations, among a grid of threads. The threads execute a special function called a "kernel". For our computations we determined that 256 threads per block is optimal by taking into account the number of registers that are used in the kernel code and the limitations on the total number of registers available in the multiprocessors. The number of blocks within a grid does not greatly impact the performance, and we decided to fix it at 16. Each Monte Carlo or quasi-Monte Carlo trajectory (sample) is computed in its own thread. If the number of samples is larger than the total number of threads (4096 in this case) we incorporate an additional cycle, launching the kernel more than once. The number of samples in practical computations should be selected approximately proportional to the number of time steps, in order to keep the balance between the stochastic and the discretization error. In this work we tested a quasi-Monte Carlo version of the algorithm, where the consecutive time steps for one sample are computed using the coordinates obtained from a suitable low-discrepancy sequence. The sampling of normally distributed random variables is achieved using the inverse function approach. Two kinds of low-discrepancy sequences were used: the Sobol' sequences with the full Owen scrambling [14], and our modification of the Halton sequences [3]. Since each time step requires three coordinates in the Kahl–Jäckel–Milstein scheme and four coordinates in the Andersen scheme, one may have to work with low-discrepancy sequences of high dimensionality. For the Sobol' sequences we were able to reach 12,500 dimensions, following the work by [11]. The generation of the Sobol' sequence is achieved by scrambling consecutive digits, using previous digits to generate "random trees" that serve to permute the next digits. This is described in detail in [5], and we basically incorporated the code from there into the program. The reason to choose the modified Halton sequences for these simulations was the theoretical result, shown in [3], that their discrepancy satisfies: $ND_N^*(\sigma) < c_s \ln^s N + O(\ln^{s-1} N)$ with a constant c_s that declines to 0 super-exponentially. The estimates for c_s in [3] are substantially better than the known estimates for the Sobol sequences and competitive with those for other low-discrepancy sequences. The reader is referred to p. 95 of the book by Niederreiter [13] for more information on the importance of these estimates. Since the constant c_s declines to 0 superexponentially, we expected better performance than the Sobol' sequences. In order to add randomness to the points being generated, we use the following definition: let $p_1, \ldots, p_s$ be prime numbers, and let $k_1, \ldots, k_s$ be "admissible" with respect to them as defined in [3]. To compute the i-th coordinate of the N-th term of the modified Halton sequence, expand N in base p_i: $N = \sum_{j=0}^{\infty} a_j p_i^j$ and set $x_N^i = \sum_{j=0}^{\infty} b_j p_i^{-j-1}$ where $b_j \equiv a_j k_i^j + c_j \mod p_i$, where a_j are c_j are fixed integers. To generate the modified Halton sequence we had to compute a list of admissible numbers for the first 16, 384 primes, because our previous codes [4] had a limitation of only 1000 coordinates. The new modifiers are available for public use from [16]. The generation of the modified Halton sequence using CUDA is based on precomputing some quantities that decrease the total number of operations required. In our preliminary tests we did not observe any advantage from using full Owen-type scrambling in the integration errors, and this is why we use this simpler version of scrambling, choosing the numbers c_j uniformly distributed among $0, \ldots, p_i - 1$. The source code for generating the modified Halton sequences using CUDA are made available under GPL at [17]. Because of the limitations of the architecture we found that in some cases we had to use the option $-maxregcount = 64$ in order to achieve correctness, although this is not always necessary, and without it one may get faster performance.

3.3 Numerical Results

We compared the performance of our algorithms with three different types of options. The underlying instrument are the shares of SPDR GLD trust, which largely follow the price of gold. We used data about American option prices to calibrate the model. In this case there is no dividend, and thus the American options are theoretically equivalent to European options. We computed three types of options: European, Asian, and barrier options. Since the FFT-based approach is far superior for European options, the first option type is used only to measure the accuracy of our methods. The barrier option that we compute is a knock-out barrier option, meaning that the payout is paid at expiration only if the price never reaches a certain threshold, which is set at 20 % above the initial price. The payout is equal to $\max(S(T) - K)$, with a strike price equal to the initial price. Our calibration yielded the following values for the parameters of the Heston model with jumps: $r = 2\,\%$, $\kappa = 1.0$, $\theta = 0.09$, $\sigma_v = 0.00001$, $\rho = 0.0$, $S_0 = 100$, $V_0 = 0.09$, $\lambda_j = 0.1$, $\mu_j = 0.1$, $\sigma_j = 0.1$, $q = 0.02$.

From Table 3.1 one can see that the scheme of Andersen is more accurate than the scheme of Kahl–Jäckel–Milstein, and from Table 3.2 that the computational time is approximately equal. In most of the computations, the modified Halton sequences outperform the Sobol' sequences with Owen scrambling in both accuracy and computational time. In some of the computations the use of the modified Halton sequences results in faster computations even compared with the standard CUDA pseudorandom generator CURAND. In any case it seems that there is a trade-off involved in the use of low-discrepancy sequences, but that they are a viable option for the two schemes

Table 3.1. Absolute errors when pricing various types of options.

		Andersen			Kahl–Jäckel–Milstein		
	Steps	*CURAND*	*Sobol*	*Halton*	*CURAND*	*Sobol*	*Halton*
European options	36	0.11	0.05	0.03	0.3	0.24	0.07
	156	0.02	0.01	0.01	0.05	0.04	0.02
Asian options	36	0.06	0.03	0.02	0.55	0.52	1.37
	156	0.02	0.01	0.01	0.38	0.36	1.17
Barrier options	36	0.08	0.04	0.01	0.17	0.23	0.03
	156	0.02	0.002	0.002	0.06	0.07	0.002

Table 3.2. Computational times in seconds.

		Andersen			Kahl–Jäckel–Milstein		
	Steps	*CURAND*	*Sobol*	*Halton*	*CURAND*	*Sobol*	*Halton*
European options	36	0.04	0.03	0.02	0.003	0.03	0.02
	156	0.07	0.56	0.3	0.04	0.81	0.41
Asian options	36	0.05	0.03	0.02	0.003	0.03	0.02
	156	0.07	0.57	0.31	0.04	0.82	0.42
Barrier options	36	0.005	0.03	0.02	0.003	0.03	0.02
	156	0.07	0.56	0.3	0.04	0.81	0.42

under consideration. We note that the codes are sufficiently general and can be used for any type of path-dependent options.

We proceed to outline a situation where fast and accurate computation of option prices is important. The calibration process for the Heston model with jumps has some inherent sources of errors, e. g., incorrect, missing or out-of-date data. Methodological difficulties arise because of unavailability of accurate spot prices, since not all options are widely traded at all times and there is substantial rounding error. This justifies the question of evaluating the sensitivity of option prices with respect to model parameters. In our study we consider the global sensitivity indices as defined by Sobol' [15]. After fitting the parameters of the model for a time period of one day, we obtain different values for the parameters. By considering the parameters of the model to be normally distributed we can determine the mean μ and the variance σ from the sampled values. Using these values we computed of the Sobol‘ sensitivity indices following the procedure from Simlab [19]. The results are shown in the Table 3.3. Looking at the numerical results we can see that in the situation under consideration the parameters, related to the jumps, have higher importance than those related o the standard Heston model and account for almost 90 % of the total variance. Since additional computational effort, required by the addition of jumps, is a small percentage of the total computational time, we can conclude that the Heston model with jumps is preferable for practical use

Table 3.3. First order (S_i) and total (S_{Ti}) sensitivity indices.

	θ_0	θ	κ	ϵ	ρ	λ_J	μ_J	v_J
μ	0.0185	0.0638	3.8466	0.6270	0.3450	0.0510	0.3190	0.0569
σ	0.0036	0.0569	6.6932	0.4436	0.0046	0.0541	0.3587	0.0276
S_i	0.01	0.12	0.06	0.07	0.0003	0.19	0.38	0.001
S_{Ti}	0.01	0.14	0.08	0.07	0.0004	0.34	0.54	0.001

and that our CUDA implementation of the Andersen scheme has the better balance of accuracy versus speed compared with the Kahl–Jäckel–Milstein scheme.

3.4 Conclusions and Future Work

The results demonstrate the efficiency of our GPGPU algorithm for the computation of option prices following the Heston model, with jumps in the underlying. The use of scrambled Sobol' sequences and especially of modified Halton sequences provides increased accuracy. The computational time is higher for the quasi-Monte Carlo algorithms, but this may not be critical in situations when more floating point computations are necessary to simulate the options' paths. Our algorithms are sufficiently general, covering the computation of any path-dependent options under the same model, and are provided under the GPL.

Acknowledgments. This work was supported part by the National Science Fund of Bulgaria under Grants DCVP 02/1, DO 02-215/2008, and the European Commission under EU FP7 project HP-SEE (under contract number 261499).

References

[1] L. B. G. Andersen, Efficient Simulation of the Heston Stochastic Volatility Model, Banc of America Securities, 2007, available online at http://ssrn.com/abstract=946405 (last accessed September 21, 2012).

[2] E. Atanassov, A New Efficient Algorithm for Generating the Scrambled Sobol' Sequence, in: *Numerical Methods and Applications*, LNCS **2**542, pp. 83–90, Springer, Berlin Heidelberg, 2003.

[3] E. I. Atanassov, On the discrepancy of the Halton sequences, *Mathematica Balkanica* **18** (2004), 15–32.

[4] E. I. Atanassov and M. K. Durchova, Generating and Testing the Modified Halton Sequences, in: I. Dimov et al. (eds.), *Numerical Methods and Applications*, pp. 91–98, LNCS 2542, Springer, Berlin Heidelberg, 2003.

[5] E. Atanassov, A. Karaivanova, and S. Ivanovska, Tuning the Generation of Sobol Sequence with Owen Scrambling, in: I. Lirkov, S. Margenov, and J. Wasniewski (eds.), *Large-Scale Scientific Computing*, pp. 459–466, LNCS 5910, Springer, Berlin Heidelberg 2010.

[6] M. Broadie and Ö. Kaya, Exact simulation of stochastic volatility and other affine jump diffusion models, *Operation Research* **54** (2006), 217–231.

[7] P. Carr and D. B. Madan, Option valuation using the Fast Fourier Transform, *Journal of Computational Finance* **2** (1999), 61–73.

[8] J. Gatheral, *The Volatility Surface: A Practitioner's Guide*, Wiley Finance, New Jersey, 2006.

[9] P. Glasserman, *Monte Carlo Methods in Financial Engineering*, Springer, New York, 2003.

[10] S. Heston, A closed-form solution for options with stochastic volatility, *Review of Financial Studies* **6** (1993), 327–343.

[11] S. Joe and F. Y. Kuo, Constructing Sobol sequences with better two-dimensional projections, *SIAM J. Sci. Comput.* **30** (2008), 2635–2654.

[12] C. Kahl and P. Jackel, Fast strong approximation Monte-Carlo schemes for SV models, *Quantitative Finance (Taylor and Francis Journals* **6** (2005), 513–536.

[13] H. Niederreiter, *Random Number Generation and Quasi-Monte Carlo Methods*, SIAM Conf. Ser. Appl. Math. 63, Society for Industrial and Applied Mathematics, 1992.

[14] A. B. Owen, Scrambling Sobo'l and Niederreiter-Xing points, *Journal of Complexity* **14** (1998), 466–489.

[15] I. M. Sobol', Global Sensitivity Indices for Nonlinear Mathematical Models and Their Monte Carlo Estimates, *Mathematics and Computers in Simulation* **55** (2001) 271–280.

[16] http://parallel.bas.bg/~emanouil/sequences/modifiers.h (last accessed September 21, 2012).

[17] Source codes for generating the modified Halton sequences using CUDA, available online at http://parallel.bas.bg/~emanouil/sequences/optprice (last accessed September 21, 2012).

[18] CUDA, http://developer.nvidia.com/category/zone/cuda-zone (last accessed September 21, 2012).

[19] SIMLAB – Sensitivity analysis, available online at http://simlab.jrc.ec.europa.eu/ (last accessed September 21, 2012).

Author information

Emanouil Atanassov, Institute of Information and Communication Technologies, Bulgarian Academy of Sciences, Sofia, Bulgaria.
Email: `emanouil@parallel.bas.bg`

Dimitar Dimitrov, Institute of Information and Communication Technologies, Bulgarian Academy of Sciences, Sofia, Bulgaria.
Email: `d.slavov@bas.bg`

Sofiya Ivanovska, Institute of Information and Communication Technologies, Bulgarian Academy of Sciences, Sofia, Bulgaria.
Email: `sofia@parallel.bas.bg`

Chapter 4

On a Game-Method for Modeling with Intuitionistic Fuzzy Estimations. Part 2

Lilija Atanassova and Krassimir Atanassov

Abstract. A new extension of Conway's Game of Life (CGL) is introduced. It is a continuation of a previous extension of CGL offered by the authors and called Game-Method for Modeling (GMM). Now we use elements of intuitionistic fuzziness that give more detailed estimations of the degrees of existence and of the non-existence of the objects occurring in the cells of the game plane. The GMM is applied over given configurations of objects in a plane. In a result of this, in a step-by-step manner, using metaheuristic rules, GMM transforms any initial configuration to a final one. If we have enough initial configurations over which to apply GMM, we can check whether the rules of GMM are statistically correct. For this case we need criteria for proximity of two configurations. Here we discuss some of these criteria. They are based on the distances between two intuitionistic fuzzy sets.

Keywords. Game-Method, Modeling, Intuitionistic Fuzzy Set.

Mathematics Subject Classification 2010. 03E72, 68Q80, 93A30.

4.1 Introduction

In a series of papers (see [6,9]), the authors extended the standard Conway's Game of Life (CGL) (see, e.g., [14]), adding to it intuitionistic fuzzy estimations (for intuitionistic fuzziness see [5]). On the other hand, more than 30 years ago the authors introduced the idea for another extension of CGL, called "game-method for modelling" (GMM) and its application in astronomy and combinatorics (see [1–3, 7, 16]). This idea found a particular application in growth and dynamics of forest inventory (see, [11, 12]) and forest fires (see [13]).

Here, using [10], we will continue with the introduction of a new extension of the standard CGL on the basis of both modifications discussed by us.

4.2 Short Remarks on the Game-Method for Modeling from Crisp Point of View

Following [10], we will mention that the standard CGL has a "universe" which is an infinite two-dimensional orthogonal grid of square cells, each of which is in one of two possible states, alive or dead, or (as an equivalent definition) in the square there is or is not an asterisk. The first situation corresponds to the case when the cell is "alive" and the second to the case when the cell is "dead".

Following the ideas from [1, 2, 7, 10], we will extend the CGL.

Let us have a set of symbols S and an n-dimensional simplex (in the sense of [15]) comprising of n-dimensional cubes (at $n = 2$, a two-dimensional net of squares). Let material points (or, for brief, objects) be found in some of the vertices of the simplex and let a set of rules A be given, containing

(1) rules for the motions of the objects along the vertices of the simplex;

(2) rules for the interactions among the objects.

Let the rules from the i-th type be marked as i-rules, where $i = 1, 2$.

When $S = \{*\}$, we obtain the standard CGL.

To each object is associated its number, n-tuple of coordinates characterizing its location in the simplex and a symbol from S reflecting the peculiarity of the object (e. g., in physical applications: mass, charge, concentration, etc.). We shall call an *initial configuration* every ordered set of $(n + 2)$-tuples with an initial component being the number of the object, the second, third, etc. until the $(n + 1)$-st being its coordinates, and the $(n + 2)$-nd being its symbol from S. We shall call a *final configuration* the ordered set of $(n + 2)$-tuples having the above form and being a result of a (fixed) initial configuration, modified during a given number of times when the rules from A have been applied.

The single application of a rule from A over a given configuration K will be called an elementary step in the transformation of the model and will be denoted by $A_1(K)$. In this sense, if K is an initial configuration, and L is a final configuration derived from K through multiple application of the rules from A, then configurations $K_0, K_1, \ldots, K_m$ will exist, for which $K_0 = K$, $K_{i+1} = A_1(K_i)$ for $0 \leq i \leq m - 1$, $K_m = L$, (the equality "=" is used in the sense of coincidence in the configurations) and this will be denoted by

$$L = A(K) \equiv A_1(A_1(\ldots A_1(K)\ldots)).$$

Let a rule P be given which juxtaposes to a combination of configurations M a single configuration $P(M)$ being the mean of the given ones. We shall call this rule a *concentrate rule.* The concentration can be made either over the values of the symbols from S for the objects or over their coordinates (not over both of them simultaneously).

Let k-th element of M ($1 \leq k \leq s$, where s is the number of elements of M) be a rectangular with $p \times q$ squares, and let the square staying on (i, j)-th place ($1 \leq i \leq p$, $1 \leq j \leq q$) contain number $d^k_{i,j} \in \{0, 1, \ldots, 9\}$. Let B be a criterion derived from physical or mathematical considerations. For two given configurations K_1 and K_2, it answers the question whether they are close to each other or not. For example, for two configurations K_1 and K_2 having the form from the above example,

$$B(K_1, K_2) = \frac{1}{p.q} \sum_{i=1}^{p} \sum_{j=1}^{q} |d_{i,j}^1 - d_{i,j}^2| < C_1$$

or

$$B(K_1, K_2) = \left(\frac{1}{p.q} \sum_{i=1}^{p} \sum_{j=1}^{q} (d_{i,j}^1 - d_{i,j}^2)^2 \right)^{\frac{1}{2}} < C_2,$$

where C_1 and C_2 are given constants.

For the set of configurations M and the set of rules A we shall define the set of configurations

$$A(M) = \{L|(\exists K \in M)(L = A(K))\}.$$

The rules A will be called statistically correct, if for a great enough (from a statistical point of view) natural number N:

$$\begin{aligned} &(\forall m > N)(\forall M = \{K_1, K_2, \ldots, K_m\}) \\ &(B(A(P(M)), P(\{L_i|L_i = A(K_i), 1 \le i \le m\})) = 1). \end{aligned} \tag{4.1}$$

The essence of the method is in the following: the set of rules A, the proximity criterion B and the concentrate rule P are fixed in advance. A set of initial configurations M is chosen and the set of the corresponding final configurations is constructed. If Equation (4.1) is valid we may assume that the rules from set A are correct within the framework of the model, i. e., they are logically consistent. Otherwise, we replace a part (or all) of them with others. If the rules become correct, then we can add to the set some new ones or transform some of the existing ones and permanently check the correctness of the newly-constructed system of rules. Thus, in consecutive steps, extending and complicating the rules in set A and checking their correctness, we construct the model of the given process. Afterwards we may check the temporal development (regarding the final system of rules A) of a particular initial configuration.

We initially check the correctness of the modeling rules, and only then do we proceed to the actual modeling. To a great extent this is due to the fact that we work over discrete objects with rules that are convenient for computer implementation. Thus, a series of checks of Equation (4.1) can be performed only to construct the configuration $A(K)$ for a given configuration K and a set of rules A.

4.3 On the Game-Method for Modeling with Intuitionistic Fuzzy Estimations

The intuitionistic fuzzy propositional calculus was introduced more than 20 years ago (see, e. g., [4, 5]). In it, if x is a variable, then its truth value is represented by the ordered couple

$$V(x) = \langle a, b \rangle$$

so that $a, b, a + b \in [0, 1]$, where a and b are the degrees of validity (existence, membership, etc., and of nonvalidity, nonexistence, etc.) of x and there the following definitions are given.

Below, we shall assume that for the two variables x and y the equalities: $V(x) = \langle a, b\rangle, V(y) = \langle c, d\rangle$ $(a, b, c, d, a + b, c + d \in [0, 1])$ hold.

The following relation of partial order is defined in IF logic (see, e. g., [5]) for every two variables x and y:

$$x \leq y \quad \text{if and only if} \quad a \leq c \quad \text{and} \quad b \geq d. \tag{4.2}$$

In [5] the theory of Intuitionistic Fuzzy Sets (IFSs) is given. Here, we provide only the definition of an IFS.

Let us have a fixed universe E and its subset A. The set

$$A^* = \{\langle x, \mu_A(x), \nu_A(x)\rangle \mid x \in E\},$$

where

$$0 \leq \mu_A(x) + \nu_A(x) \leq 1$$

is called IFS and functions $\mu_A : E \to [0, 1]$ and $\nu_A : E \to [0, 1]$ represent *degree of membership (validity, etc.)* and *non-membership (non-validity,* etc.).

For brevity, we shall write below A instead of A^*.

In [8] some intuitionistic fuzzy parametric distances between two IFSs are discussed. Below we list ten of them that will use in the present research.

Let everywhere below $0 \leq \varepsilon \leq \eta < 1$ be fixed. The ten distances between IFSs, introduced in [8] are

$$\begin{aligned} d'_{\varepsilon,\eta;str\_opt}(A, B)(x) = \big\langle &\min(\mu_A(x), \nu_B(x) + \varepsilon) + \min(\mu_B(x), \nu_A(x) + \varepsilon) \\ &- \min(\mu_A(x), \nu_B(x) + \varepsilon).\min(\mu_B(x), \nu_A(x) + \varepsilon), \\ &\max(\nu_A(x), \mu_B(x) - \eta).\max(\nu_B(x), \mu_A(x) - \eta)\big\rangle, \end{aligned}$$

$$\begin{aligned} d'_{\varepsilon,\eta;opt}(A, B)(x) = \big\langle &\max(\min(\mu_A(x), \nu_B(x) + \varepsilon), \min(\mu_B(x), \nu_A(x) + \varepsilon)), \\ &\min(\max(\nu_A(x), \mu_B(x) - \eta), \max(\nu_B(x), \mu_A(x) - \eta))\big\rangle, \end{aligned}$$

$$\begin{aligned} d'_{\varepsilon,\eta;aver}(A, B)(x) = \Big\langle &\frac{\min(\mu_A(x), \nu_B(x) + \varepsilon) + \min(\mu_B(x), \nu_A(x) + \varepsilon)}{2}, \\ &\frac{\max(\nu_A(x), \mu_B(x) - \eta) + \max(\nu_B(x), \mu_A(x) - \eta)}{2}\Big\rangle, \end{aligned}$$

$$\begin{aligned} d'_{\varepsilon,\eta;pes}(A, B)(x) = \big\langle &\min(\min(\mu_A(x), \nu_B(x) + \varepsilon), \min(\mu_B(x), \nu_A(x) + \varepsilon)), \\ &\max(\max(\nu_A(x), \mu_B(x) - \eta), \max(\nu_B(x), \mu_A(x) - \eta))\big\rangle, \end{aligned}$$

$$d'_{\varepsilon,\eta;str\_pes}(A,B)(x) = \langle \min(\mu_A(x), \nu_B(x)+\varepsilon).\min(\mu_B(x), \nu_A(x)+\varepsilon),$$
$$\max(\nu_A(x), \mu_B(x)-\eta) + \max(\nu_B(x), \mu_A(x)-\eta)$$
$$- \max(\nu_A(x), \mu_B(x)-\eta).\max(\nu_B(x), \mu_A(x)-\eta)\rangle.$$

$$d''_{\varepsilon,\eta;str\_opt}(A,B)(x) = \langle \min(\max(\varepsilon, \mu_A(x)-\eta+\varepsilon), \nu_B(x)+\varepsilon)$$
$$+ \min(\max(\varepsilon, \mu_B(x)-\eta+\varepsilon), \nu_A(x)+\varepsilon)$$
$$- \min(\max(\varepsilon, \mu_A(x)-\eta+\varepsilon),$$
$$\nu_B(x)+\varepsilon).\min(\max(\varepsilon, \mu_B(x)-\eta+\varepsilon), \nu_A(x)+\varepsilon),$$
$$\max(0, \min(1-\eta, \nu_A(x)+\varepsilon-\eta), \mu_B(x)-\eta)$$
$$.\max(0, \min(1-\eta, \nu_B(x)+\varepsilon-\eta), \mu_A(x)-\eta)\rangle,$$

$$d''_{\varepsilon,\eta;opt}(A,B)(x) = \langle \max(\min(\max(\varepsilon, \mu_A(x)-\eta+\varepsilon), \nu_B(x)+\varepsilon),$$
$$\min(\max(\varepsilon, \mu_B(x)-\eta+\varepsilon), \nu_A(x)+\varepsilon)),$$
$$\min(\max(0, \min(1-\eta, \nu_A(x)+\varepsilon-\eta), \mu_B(x)-\eta),$$
$$\max(0, \min(1-\eta, \nu_B(x)+\varepsilon-\eta), \mu_A(x)-\eta))\rangle,$$

$$d''_{\varepsilon,\eta;aver}(A,B)(x) =$$
$$\langle \frac{\min(\max(\varepsilon,\mu_A(x)-\eta+\varepsilon),\nu_B(x)+\varepsilon)+\min(\max(\varepsilon,\mu_B(x)-\eta+\varepsilon),\nu_A(x)+\varepsilon)}{2},$$
$$\frac{\max(0,\min(1-\eta,\nu_A(x)+\varepsilon-\eta),\mu_B(x)-\eta)+\max(0,\min(1-\eta,\nu_B(x)+\varepsilon-\eta),\mu_A(x)-\eta)}{2}\rangle,$$

$$d''_{\varepsilon,\eta;pes}(A,B)(x) = \langle \min(\max(\varepsilon, \mu_A(x)-\eta+\varepsilon, \nu_B(x)+\varepsilon),$$
$$\max(\varepsilon, \mu_B(x)-\eta+\varepsilon), \nu_A(x)+\varepsilon),$$
$$\max(0, \min(1-\eta, \nu_A(x)+\varepsilon-\eta), \mu_B(x)-\eta,$$
$$\min(1-\eta, \nu_B(x)+\varepsilon-\eta), \mu_A(x)-\eta)\rangle,$$

$$d''_{\varepsilon,\eta;str\_pes}(A,B)(x) = \langle \min(\max(\varepsilon, \mu_A(x)-\eta+\varepsilon), \nu_B(x)+\varepsilon)$$
$$.\min(\max(\varepsilon, \mu_B(x)-\eta+\varepsilon), \nu_A(x)+\varepsilon),$$
$$\max(0, \min(1-\eta, \nu_A(x)+\varepsilon-\eta), \mu_B(x)-\eta)$$
$$+ \max(0, \min(1-\eta, \nu_B(x)+\varepsilon-\eta), \mu_A(x)-\eta)$$
$$\max(0, \min(1-\eta, \nu_A(x)+\varepsilon-\eta), \mu_B(x)-\eta)$$
$$.\max(0, \min(1-\eta, \nu_B(x)+\varepsilon-\eta), \mu_A(x)-\eta)\rangle.$$

We must mention that these distances have the forms of intuitionistic fuzzy pairs. These distances conditionally were called, respectively, "strong optimistic", "optimistic", "average", "pessimistic", and "strong pessimistic".

4.4 Main Results

As we saw above, each asterisk from CGL corresponds to a symbol in the GMM. Now, following ideas from and continuing [10], we will perform the next step of extension. For this step, we have two possibilities which are equivalent in the general case.

First, we can change the symbols from S, with which the objects in GMM are marked, to IF-couples, determining the degree of existence and degree of non-existence of this object.

Second, we can keep the symbols from S and attach the same IF-couple to them.

Let us discuss the second case, because, obviously, we can obtain the first one directly.

As a first step of the research in the present direction, it is convenient to assume that the objects will not change their IF-parameters as a result of movement from one cell to another that has a common side with the former cell. In future research, we will discuss the more complex situation, when in the result of the movement the IF-parameters changes (e. g., decreases). Therefore, the criteria for existence of an object before and after its movement will be the same, and we can use the conditions from [9], but in [10] we extended its criteria list.

Now we will give intuitionistic fuzzy interpretations of these of the GMM-elements which have not yet been discussed.

Let us assume that the square $\langle i, j\rangle$ is assigned a pair of real numbers $\langle \mu_{i,j}, \nu_{i,j}\rangle$, so that $\mu_{i,j} + \nu_{i,j} \leq 1$. We can call the numbers $\mu_{i,j}$ and $\nu_{i,j}$ degree of existence and degree of nonexistence of an object, or (in CGL and its IF-extension), of a symbol "*" in square $\langle i, j\rangle$. Therefore, $\pi(i, j) = 1-\mu_{i,j}-\nu_{i,j} \leq 1$ will correspond to the degree of uncertainty, e. g., lack of information about existence of an asterisk in the respective cell. In [10] we formulated seven criteria for the existence of an object in a cell which will include as a particular case the standard game. From these criteria it follows that if one is valid – let it be the s-th criterion ($1 \leq s \leq 7$) – then we can assert that the object exists with respect to the s-th criterion and it will therefore exist with respect to all other criteria, the validity of which follows from the validity of the s-th criterion.

On the other hand, if s-th criterion is not valid, then we can say that the object does not exist with respect to the s-th criterion. It is very important that in this case the square may not be totally empty. We can tell that the square is *"s-full"* if it contains an object with respect to the s-th criterion, or that it is *"s-empty"* if it is empty or contains an object not satisfying the s-th criterion.

For the aims of the GMM, it will be suitable to use (with respect to the type of the concrete model) one of the first four criteria for the existence of an object. Let us say that for each fixed square $\langle i, j\rangle$ there is an object by the s-th criterion ($1 \leq s \leq 4$), whether this criterion confirms the existence of the object.

For the aims of the present chapter, we can put $\varepsilon = \eta = 0$ in the formulas of distances between IFSs and these distances will obtain the form

$$\begin{aligned} d_{str\_opt}(A, B)(x) &= d'_{str\_opt}(A, B)(x) = d''_{str\_opt}(A, B)(x) \\ &= \big\langle \min(\mu_A(x), \nu_B(x)) + \min(\mu_B(x), \nu_A(x)) \\ &\quad - \min(\mu_A(x), \nu_B(x)). \min(\mu_B(x), \nu_A(x)), \\ &\quad \max(\nu_A(x), \mu_B(x)). \max(\nu_B(x), \mu_A(x))\big\rangle, \end{aligned}$$

$$\begin{aligned} d_{opt}(A, B)(x) &= d'_{opt}(A, B)(x) \\ &= d''_{opt}(A, B)(x) \\ &= \big\langle \max(\min(\mu_A(x), \nu_B(x)), \min(\mu_B(x), \nu_A(x))), \\ &\quad \min(\max(\nu_A(x), \mu_B(x)), \max(\nu_B(x), \mu_A(x)))\big\rangle, \end{aligned}$$

$$\begin{aligned} d_{aver}(A, B)(x) &= d'_{aver}(A, B)(x) \\ &= d''_{aver}(A, B)(x) \\ &= \Big\langle \frac{\min(\mu_A(x), \nu_B(x)) + \min(\mu_B(x), \nu_A(x))}{2}, \\ &\quad \frac{\max(\nu_A(x), \mu_B(x)) + \max(\nu_B(x), \mu_A(x))}{2}\Big\rangle, \end{aligned}$$

$$\begin{aligned} d_{pes}(A, B)(x) &= d'_{pes}(A, B)(x) \\ &= d''_{pes}(A, B)(x) \\ &= \big\langle \min(\min(\mu_A(x), \nu_B(x)), \min(\mu_B(x), \nu_A(x))), \\ &\quad \max(\max(\nu_A(x), \mu_B(x)), \max(\nu_B(x), \mu_A(x)))\big\rangle, \end{aligned}$$

$$\begin{aligned} d_{str\_pes}(A, B)(x) &= d'_{str\_pes}(A, B)(x) = d''_{str\_pes}(A, B)(x) \\ &= \big\langle \min(\mu_A(x), \nu_B(x)). \min(\mu_B(x), \nu_A(x)), \\ &\quad \max(\nu_A(x), \mu_B(x)) + \max(\nu_B(x), \mu_A(x)) \\ &\quad - \max(\nu_A(x), \mu_B(x)). \max(\nu_B(x), \mu_A(x))\big\rangle. \end{aligned}$$

Therefore, the symbols for prime and second can be omitted.

The following assertion can be proved on the basis of above definitions of distances and inequality (4.2).

Theorem 4.1. *For every two IFSs A and B and for every $x \in E$:*

$$\begin{aligned} d_{str\_pes}(A, B)(x) &\le d_{pes}(A, B)(x) \le d_{aver}(A, B)(x) \\ &\le d_{opt}(A, B)(x) \le d_{str\_opt}(A, B)(x). \end{aligned}$$

Now we see that, each GMM-configuration K can be interpreted as an IFS

$$K = \{\langle\langle i, j\rangle, \mu_{i,j}, \nu_{i,j}\rangle | 1 \le i \le p \;\&\; 1 \le j \le q\}.$$

Each of the above distances can be used as a criterion B for proximity between two configurations K_1 and K_2, so that symbols K_1 and K_2 will stay on places of A and B in the formulas of distances. We therefore obtain intuitionistic fuzzy interpretation of the criterion for distances for GMM.

About rule P, the list of the possibilities for intuitionistic fuzzy interpretations is not long: if we use interpretations that are analogous to "strong optimistic", "optimistic", "pessimistic", or "strong pessimistic" distances, we will change the sense of the model in some (μ- or ν-) direction. Only the analogue of the "average" distance is suitable for our aims. In this case, we must use the following operations for the two IFSs A and B(see [5]).

$$A@B = \left\{\left\langle x, \frac{\mu_A(x) + \mu_B(x)}{2}, \frac{\nu_A(x) + \nu_B(x)}{2}\right\rangle | x \in E\right\}$$

and for n IFSs $A_1, A_2, \ldots, A_n$:

$$\underset{i=1}{\overset{n}{@}} A_i = \left\{\left\langle x, \sum_{i=1}^{n} \frac{\mu_{A_i}(x)}{n}, \sum_{i=1}^{n} \frac{\nu_{A_i}(x)}{n}\right\rangle | x \in E\right\}.$$

In particular,

$$A_1 @ A_2 = \underset{i=1}{\overset{2}{@}} A_i.$$

Therefore, if we have a set of (initial or final) configurations M, we can construct the average configuration

$$P(M) = \underset{i=1}{\overset{n}{@}} K_i.$$

4.5 Conclusion

In future research in this direction on the extension of CGL, we will discuss the possibility for change of the IF-parameters of the objects in the plane, as a result of their movement from one cell to another, which has a common side with the previous cell.

Acknowledgments. The second author is grateful for the support provided by the projects DID-02-29 "Modeling processes with fixed development rules" and BIn-2/09 "Design and development of intuitionistic fuzzy logic tools in information technologies", funded by the National Science Fund, Bulgarian Ministry of Education, Youth and Science.

References

[1] K. Atanassov, On a combinatorial game-method for modeling, *Advances in Modeling & Analysis* **19** (1994), pp. 41–47.

[2] K. Atanassov, Application of a combinatorial game-method in combinatorial geometry. Part 1: Combinatorial algorithms for solving variants of the Steiner-Rosenbaum's problem, *Advances in Modeling & Analysis* **2**(1–2) (1998), 23–29.

[3] K. Atanassov, Application of a combinatorial game-method in combinatorial geometry. Part 2: Algorithm for grouping and transferring of points and a general algorithm, *Advances in Modelling & Analysis* **2**(1–2) (1998), 31–36.

[4] K. Atanassov, Two variants of intuitonistic fuzzy propositional calculus. Preprint IM-MFAIS-5–88, Sofia, 1988.

[5] K. Atanassov, *Intuitionistic Fuzzy Sets*, Springer Physica-Verlag, Heidelberg, 1999.

[6] K. Atanassov, L. Atanassova, A game-method for modeling, in: *Proc. of the 3rd International School "Automation and Scientific Instrumentation"*, Varna, Oct. 1984, pp. 229–232.

[7] K. Atanassov, L. Atanassova, and D. Sasselov, On the combinatorial game-method for modeling in astronomy, *Comptes Rendus de l'Academie Bulgare des Sciences* **47**(9) (1994), 5–7.

[8] K. Atanassov, A new approach to the distances between intuitionistic fuzzy sets, in: E. Hullermeier, R. Kruse and F. Hoffmann (eds.), *Information Processing and Management of Uncertainty in Knowledge-Based Systems*, pp. 581–590, Springer, Heidelberg, 2010.

[9] L. Atanassova and K. Atanassov, On a game-method for modeling with intuitionistic fuzzy estimations. Part 1, 8th International Conference LSSC'11, Sozopol, June 6–10, 2011 (in press).

[10] L. Atanassova and K. Atanassov, *Intuitionistic fuzzy interpretations of Conway's game of life*, pp. 232–239, Lecture Notes in Computer Science 6046, Springer, Berlin Heidelberg, 2011.

[11] D. Dimitrov, Modeling the growth and dynamics of forest stands by game-method, *Advances in Modeling & Analysis* **2** (1998), 11–22.

[12] D. Dimitrov, Modeling the growth and dynamics of forest stands by extended game-method, *Advances in Modeling & Analysis* **4** (1999), 7–21.

[13] N. Dobrinkova, S. Fidanova, and K. Atanassov. *Game-method model for field fires*, pp. 173–179, Lecture Notes in Computer Science 5910, Springer, Berlin Heidelberg, 2010.

[14] http://en.wikipedia.org/wiki/Conway%27s_Game_of_Life, 2012 (last accessed September 21, 2012).

[15] K. Kuratovski, *Topology*, Academic Press, New York, 1966.

[16] D. Sasselov and K. Atanassov, On the generalized nets realization of a combinatorial game-method for modelling in astronomy, *Advances in Modeling & Analysis* **23** (1995), 59–64.

Author information

Lilija Atanassova, Institute of Information and Communication Technologies, Bulgarian Academy of Sciences, Sofia, Bulgaria.
Email: l.c.atanassova@gmail.com

Krassimir Atanassov, Institute of Biophysics and Biomedical Engineering, Bulgarian Academy of Sciences, Sofia, Bulgaria.
Email: krat@bas.bg

Chapter 5

Generalized Nets, ACO Algorithms, and Genetic Algorithms

Vassia Atanassova, Stefka Fidanova, Ivan Popchev, and Panagiotis Chountas

Abstract. In real life as well as in industry there arise difficult problems which need a huge amount of computational recourses. It is unpractical to solve these kinds of problems with exact or traditional numerical methods. Therefore metaheuristic methods which can quickly find good solutions are applied. Sometimes the metaheuristic methods fall into local minimum and stop. To avoid this or to secure the process hybrid methods are applied which are a combination of some metaheuristic and exact methods or local search procedures. In this chapter we combine Ant Colony Optimization (ACO) with Genetic Algorithm (GA), which are both metaheuristic methods. The hybridization is in two directions: from ACO to GA and from GA to ACO. We describe the new hybrid algorithm with Generalized Net (GN), which helps us to see the processes in detail and to improve them.

Keywords. Generalized Nets, Ant Colony Optimization, Genetic Algorithm.

Mathematics Subject Classification 2010. 68T20, 93A30.

5.1 Introduction

There are countless applications for optimization. Every process has a potential of being optimized. Indeed many challenging applications in science and industry can be formulated as optimization problems. A large number of real -ife optimization problems in science, engineering, economics, and business are complex and difficult to solve. They cannot be solved by exact algorithms or traditional numerical methods in a reasonable time frame. In such cases so-called metaheuristic methods are applied.

Metaheuristics are increasingly popular in different research areas and industries. Most of them mimic natural metaphors to solve complex optimization problems (evolution, ant colony, particle swarm, immune system). Unlike exact methods, metaheuristics allow us to tackle large-scale problem instances by delivering satisfactory solutions within a reasonable amount of time. Their use in many applications shows their efficiency and effectiveness for solving large and complex problems. Among the most successive metaheuristics are Genetic Algorithm and Ant Colony Optimization.

Generalized nets are a tool for describing algorithms and processes. They help to see details and to improve algorithms. The concept of generalized nets was defined about 25 years ago as an extension of Petri nets and other modifications of them. It was proved that for each of the individual classes of Petri nets it is possible to construct a particular GN, describing the way of functioning and the results of the work of each net from this class. In general terms, the Petri net is a dichromatic directed graph in which points conventionally called "tokens" are moving. They mark the development of a process, the step of which are represented by the vertices of the graph. GNs' being a real extension of the Petri nets is expressed by the fact that the GN tokens enter the net (whose graphical structure is again represented by a graph) with assigned initial characteristics, and during their movement in the net the tokens obtain new characteristic while keeping the old ones. Thus the tokens are transformed into "individuals" with their own history record. The condition from predicate-transition nets, related to the movement of the tokens, is replaced here with a whole predicate matrix that allows a much more detailed description of the causes for movement in the different possible directions. In [2,4], GNs without some of their components were defined and thus named reduced GN, as well as more than 20 different types of extensions of GN. Various aspects of the theory of GN were described; algebraic, topological, logical, operational, programming, and others. Within the operational aspect, six types of operators were defined which can modify the structure or the behavior of a given GN in a user-defined way.

The GN model of the hybrid ACO-GA metaheuristics gives us a detailed notation and better understanding of the algorithm behavior. This representation can suggest ideas for development and improvement of the method. Our aim is to help the optimization algorithm developer to understand in detail and, if possible, improve the behavior of their algorithms. Our novelty is the hybrid ACO-GA algorithm and its representation by GN.

The organization of the paper is as follows: In Section 5.2 we briefly introduce the ACO and GA. In Section 5.3 we propose a hybrid ACO-GA algorithm and describe it with GN. At the end we give some conclusions and directions for future work.

5.2 ACO and GA

Genetic Algorithms are considered a promising novel concept. They represent a parallel, global search technique that emulates natural genetic operators [13]. They are stochastic search methods designed for exploring complex problem space in order to find optimal solutions using minimal information on the problem to guide the search. GAs are characterized by the use of a population of multiple structures (individuals) to perform the searches throughout many different areas of the problem space at the same time. The individuals encode tentative solutions which are manipulated competitively by applying to them some stochastic operators in order to find a satisfactory solution. A GA proceeds in an iterative way by generating a new population of individuals from a previous one. In a traditional GA the initial population is generated randomly.

A fitness function associates a value to every individual which is meaningful for its suitability to the problem at hand. The canonical algorithm applies stochastic operators, such as selection, crossover, and mutation, on a population, in order to compute a whole generation of new individuals. The stopping criterion is usually set as reaching a preprogrammed number of iterations of the algorithm and/or to find an individual with a given error if the optimum or an approximation of it is known beforehand. Since it simultaneously evaluates many points in the parameter space, the algorithm is more likely to converge towards the global solution. A GA requires only information concerning the quality of the solution produced by each parameter set. Since GAs do not require any problem-specific information, they are more flexible and adaptable than most search methods.

Until now, using the apparatus of generalized nets, a few GN models regarding genetic algorithm performance have been developed. The first GN model describes the genetic algorithm search procedure, [15]. The GN model simultaneously evaluates several fitness functions, ranks the individuals according to their fitness, and has the opportunity to choose the best fitness function regarding the specific problem domain. The second GN model describes a selection of genetic algorithm operators [5]. The GN model has the possibility to test different groups of the defined genetic algorithm operators and to choose the most appropriate combination of them. The developed GN executes a genetic algorithm and implements tuning the genetic operators, as well as the fitness function, with regard to the considered problem.

The Ant Colony Optimization is a new metaheuristic method. ACO algorithms are applied to real-life and industrial problems for which sufficiently good solutions are required in a short time. ACO finds good solutions for optimization problems with restrictive constraints, such as the multiple knapsack problem. The method represents a multiagent system, where low-level interaction between single agents results in complex behavior of the whole ant colony. As mentioned above, it imitates the behavior exhibited by real ants when searching for food. Ants are social insects who live in colonies and whose behavior is aimed more at the survival of the colony as a whole rather than of a single individual in the colony. An important and interesting aspect of ant colonies is how ants find the shortest path between the food source and their formicary. Ants communicate information about food sources by means of a chemical substance called pheromone, which they secrete as they move along. Analogously, ACO is based on the indirect communication of a colony of simple agents, called "artificial ants", mediated by "artificial pheromone" trails. The pheromone trails in ACO algorithms are represented by distributed numerical information, which ants are using to probabilistically construct solutions to the problem and which is being adapted by the ants during the algorithm's execution to reflect their search experience. Artificial ants not only imitate the behavior described, but they also apply additional problem-specific heuristic information. The idea is developed by Moyson in [3]. The first example of ant algorithm is Ant System [1] which has been applied to finding solutions to various difficult combinatorial optimization problems. As of today, different versions of elaborated ACO algorithms, such as the ant colony system [14], the ant algorithm with elitist ants [6], the max-min ant system [7], the ant algorithm with additional reinforcement [9], and the best-worst ant system [8], have been applied to many optimization problems.

To date, several GN models describing the functioning and the results of the work of ACO algorithms have been prepared [10–12].

5.3 GN for Hybrid ACO-GA Algorithm

Metaheuristics are iterative methods that are applied with the aim of finding a solution which meets the customer's/user's requirements within a reasonable amount of computation time. In contrast to many classical methods, metaheuristics do not build a model of the tackled optimization problem, but rather treat the problem as it is (black-box optimization). They are therefore directly applicable to complex real-world problems with relatively few modifications. In recent years it has become evident that the concentration on a sole metaheuristic is rather restrictive. A skilled combination of a metaheuristic with other optimization techniques, a so called hybrid metaheuristic, can provide a more efficient behavior and a higher flexibility when dealing with real-world and large-scale problems. In general, hybrid metaheuristic approaches can be classified as either "collaborative combinations" or "integrative combinations". Collaborative combinations are based on the exchange of information between a metaheuristic and another optimization technique running sequentially (or in parallel). Integrative combinations utilize another optimization technique (e. g., dynamic programming) as a subordinate part of a metaheuristic. For the above-mentioned reasons, hybrid metaheuristics currently enjoy an increasing interest within the optimization community.

Most authors combine some metaheuristics with local search procedure or with some exact method. Our idea is to combine two metaheuristics, ACO and GA. ACO is a constructive method which does not need initial solutions. GA is a population-based method and in traditional GA initial population is randomly generated. In this random generation the initial solutions can be very far from the optimal solutions and need a lot of iterations to get close to it. Therefore our idea is to generate initial solutions by ACO which are used afterwards as an initial population in GA. Thus the GA will start with the population which is closer to the optimal solution. After a number of iterations, the GA go to stagnation, and the population ceases to be improved. So the next idea is to return the GA solutions to the ACO algorithm and to update the pheromone accordingly, after which the ACO is run with updated pheromone, thus generating a new population for GA. Thus we will hybridize in two directions: GA with ACO and ACO with GA. We can use any variant of ACO and GA, depending on the class of problems to which the algorithm is applied. We describe the ACO and the GA with GNs (G_{ACO} and G_{GA} respectively), and using them we prepare a GN describing the hybrid ACO-GA algorithm. The problem is described with G_{proc}.

Each one of the GNs G_{ACO} and G_{GA} has one input and one output place. Let them be, respectively, $i_{ACO}, o_{ACO}, i_{GA}, o_{GA}$.

Let token α of GN G_{proc} enter place l_1 of GN with initial characteristic

"current problem description (graph of the problem, problem constraints, etc.".

$$Z_1 =< \{l_1, o_{GA}\}, \{l_2, l_3\}, \begin{array}{c|cc} & l_2 & l_3 \\ \hline l_1 & true & false \\ o_{GA} & W_{GA,2} & W_{GA,3} \end{array} ,$$

where
$W_{GA,2}$ = "a next iteration is necessary",
$W_{GA,3} = \neg W_{GA,2}$,
$\neg P$ is the negation of predicate P.

Token α from place l_1 enters place l_2 with the new characteristic

initialization of the graph of the problem (initial pheromonne).

When the truth-value of $W_{GA,2}$ is *true*, token α from place o_{GA} enters place l_2 with the characteristic

pheromonne updating according current solutions,

while it enters place l_3 without the new characteristic

$$Z_2 =< \{l_2, l_5\}, \{i_{ACO}\}, \begin{array}{c|c} & i_{ACO} \\ \hline l_2 & true \\ l_5 & true \end{array} .$$

The α-tokens from places l_2 or l_5 enter place i_{ACO} without a new characteristic. It transfers through GN G_{ACO}, and on the way out it (through place o_{ACO}) obtains the characteristic

current solutions of ACO-algorithm (population generations).

$$Z_3 =< \{o_{ACO}\}, \{l_4, l_5\}, \begin{array}{c|cc} & l_4 & l_5 \\ \hline o_{ACO} & W_{ACO,4} & W_{ACO,5} \end{array} ,$$

where
$W_{ACO,4}$ = *the end-condition is satisfied,*
$W_{ACO,5} = \neg W_{ACO,4}$.

When the truth-value of $W_{ACO,4}$ is *true*, token α enters place l_4 with the characteristic

representation of the current solutions (populations) in a propriate form of the GA,

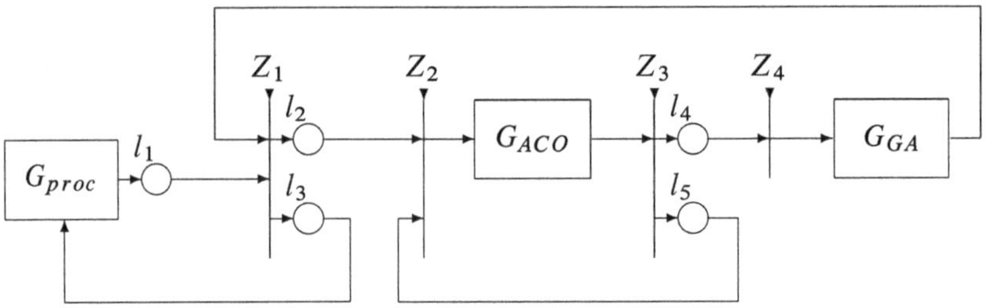

Figure 5.1. Generalized Net (GN).

while it enters place l_5 without a new characteristic.

$$Z_4 =< \{l_4\}, \{i_{GA}\}, \begin{array}{c|c} & l_5 \\ \hline l_4 & true \end{array} >.$$

Token α from place l_4 enters place i_{GA} with the characteristic

current population (solutions) of the GA.

5.4 Conclusion

In this chapter we propose new hybrid metaheuristic which is a combination of two metaheuristics, ACO and GA. The hybridization is in two directions, ACO to GA and GA to ACO. We hope that the new hybrid metaheuristic will combine the advantages of the two composite metaheuristics. Any of the ACO and the GA algorithms can be applied, depending on which is most appropriate for soling the current problem. We describe our new algorithm with GN. The aim is to see in detail the algorithm behavior and the possibilities for its improvement.

Acknowledgments. The authors would like to acknowledge the support provided by the Bulgarian National Science Fund under Grants Ref. No. DID-02-29 “Modeling Processes with Fixed Development Rules” and DTK-02-44 “Efficient Monte Carlo Methods for Large-Scale Scientific Problems”.

References

[1] J. Alexieva, E. Choy, and E. Koychev, Review and bibloigraphy on generalized nets theory and applications, in: E. Choy, M. Krawczak, A. Shannon and E. Szmidt, (eds), *A Survey of Generalized Nets*, +pp. 207–301, Raffles KvB Monograph 10, 2007.

[2] K. Atanassov, *Generalized Nets*, World Scientific, Singapore, New Jersey, London, 1991.

[3] K. Atanassov, *Generalized Nets and Systems Theory*, "Prof. M. Drinov" Academic Publishing House, Sofia, 1997.

[4] K. Atanassov, *On Generalized Nets Theory*, Prof. M. Drinov Publishing House, Sofia, 2007.

[5] I. T. Dimov, B. Philippe, A. Karaivanova, and C. Weihrauch, Robustness and applicability of Markov chain Monte Carlo algorithms for eigenvalue problems, *Applied Mathematical Modelling* **32**(8) (2007), 1511–1529, available online at http://www.sciencedirect.com/science/journal/0307904X/32/8, 2012 (last accessed September 21, 2012).

[6] M. Dorigo, V. Maniezzo, and A. Colorni, The Ant System: Optimization by a Colony of Cooperative Agents, *IEEE Transactions on Systems, Man and Cybernetics B* **26**(1) (1996), 29–41.

[7] M. Dorigo and L. M. Gambardella, Ant Colony System: A Cooperative Learning Approach to the Traveling Salesman Problem, *IEEE Transactions on Evolutionary Computing* **1** (1999), 53–66.

[8] M. Dorigo, G. Di Caro, and M. Gambardella, Ant Algorithms for Distributed Discrete Optimization, *J. Artificial Life* **5** (1999), 137–172.

[9] S. Fidanova., ACO Algorithm with Additional Reinforcement, in: M. Dorigo, G. Di Caro, and M. Sampels (eds.), *From Ant Colonies to Artificial Ants*, pp. 292–293, Lecture Notes in Computer Science 2463, Springer, Berlin Heidelberg, 2003.

[10] Fidanova, S. and Atanassov, K. Generalized net models of the process of ant colony optimization with intuitionistic fuzzy estimations, in: K, Atanassov and A. Shannon (eds.) *Proceedings of the Ninth International Workshop on Generalized Nets*, Sofia, July 4, 2008, vol. 1, pp. 41–48, 2008.

[11] S. Fidanova and K. Atanassov, Generalized net models for the process of hybrid ant colony optimization, *Comptes Rendus de l'Academie bulgare des Sciences* **61**(12), (2008), 1535–1540.

[12] S. Fidanova and K. Atanassov, *Generalized nets as tools for modelling of the ant colony optimization algorithms*, pp. 326–333, Lecture Notes in Computer Science 5910, Springer Berlin Heidelberg, 2010.

[13] D. Goldberg, *Genetic Algorithms in Search, Optimization and Machine Learning*. Addison-Weslcy, Reading, MA, 1989.

[14] F. Moyson and B. Manderick, The Collective Behavior of Ants: An Example of Self-Organization in Massive Parallelization, in: *Proceedings of AAAI Spring Symposium on Parallel Models of Intelligence*, Stanford, CA, 1988

[15] O. Roeva, K. Atanassov, and A. Shannon, Generalized net for evaluation of the genetic algorithm fitness function, in: *Proceedings of the 8th Int. Workshop on Generalized Nets*, Sofia, pp. 48–55, 2007.

Author information

Vassia Atanassova, Institute of Information and Communication Technologies, Bulgarian Academy of Sciences, Sofia, Bulgaria.
Email: vassia.atanassova@gmail.com

Stefka Fidanova, Institute of Information and Communication Technologies, Bulgarian Academy of Sciences, Sofia, Bulgaria.
Email: stefka@parallel.bas.bg

Ivan Popchev, Institute of Information and Communication Technologies, Bulgarian Academy of Sciences, Sofia, Bulgaria.
Email: ipopchev@iit.bas.bg

Panagiotis Chountas, Department of Business Information Systems, School of Electronics & Computer Science, University of Westminster, London, UK.
Email: p.i.chountas@westminster.ac.uk

Chapter 6

Bias Evaluation and Reduction for Sample-Path Optimization

Fabian Bastin

Abstract. Stochastic programming allows capturing uncertainties by incorporating random variables into the optimization problem. While this makes the model more realistic, it also leads to various computational issues. In the context of the expected value optimization, different techniques have been proposed, in particular stochastic gradient and sample-path optimization. We focus here on the latter case, assuming, however, that the expectation operator can itself be embedded as an argument of another function. This creates a simulation bias that can be partly corrected, but care must be exercised due to the additional presence of optimization bias. While this optimization bias is well known within the stochastic programming community, it is often dismissed in some practical applications. This lack of understanding can lead to unexpected results, especially as the quality of the optimal solution can decrease when the simulation bias is corrected. We explore the interaction of these two bias origins and numerically illustrate it in the context of simulated maximum likelihood.

Keywords. Stochastic Programming, Nonlinear Programming, Simulation Bias, Optimization Bias, Bias Reduction, Mixed Logit.

Mathematics Subject Classification 2010. 90C15, 90C30, 90C59, 62L20.

6.1 Introduction

A classical approach for solving stochastic optimization programs involving expectations is to replace them by some Monte Carlo (or quasi-Monte Carlo) approximations leading to methods known as sample-path optimization or optimization of sample average approximations (see, e. g., [15, 16]). While most of the considered objective functions consist of a single expectation or a linear combination of expectations, we can face a nonlinear transformation of these expectations, so that any sample approximation will typically produce some bias.

A specific example is the estimation based on simulated log-likelihood maximization. The idea of correcting the bias of such a log-likelihood function is not new and can be found, e. g., in Gouriéroux and Monfort (1996), who derived a similar expression using different arguments. They did not, however, report numerical experiments, nor did they discuss the interaction with optimization bias, while statistical properties of simulated log-likelihood estimators have been explored for some time. We refer to

Gouriéroux and Monfort [9, Chap. 3], for a coverage of them. In [8], Gouriéroux and Monfort express the simulation bias in terms of the information matrix and proposed a first-order correction based on a limited development of the logarithm function. The concern about this simulation bias has increased during recent years in the estimation literature, with some properties rediscovered by (mostly applied) researchers. For instance, Tsagkanos [19] suggests using a bootstrap bias estimate. In [2], Bastin, Cirillo, and Toint proposed an adaptive Monte Carlo algorithm for nonlinear stochastic programming and applied it in the context of mixed logit models, the estimation of which can be performed by maximizing a simulated log-likehood function. They take the bias of the log-likelihood into account and give a bias estimator similar to the one proposed by Gouriéroux and Monfort. The only difference can be translated by the use of the unbiased variance estimator in [2] and the (slightly) biased variance estimator in [8] as the sum of squares is divided by the number of observations. Bastin and Cirillo [1] later used this Taylor-based estimator as an attempt to improve the computed estimators for mixed-logit models. More recently, Kristensen and Salanie [10] made a comparison between bootstrap, Taylor, and a new method based on Newton–Raphson, still in the context of mixed logit models. Kristensen and Salanie only consider synthetic data, and while the Newton–Raphson estimator gave the best correction, they recommended the Taylor-based correction for practical purposes, as this bias estimation remains close to the Newton–Raphson estimator, and can be obtained at a much cheaper price. In contrast, numerical results obtained by Bastin and Cirillo were more mitigated, as the solution quality after applying the correction was sometimes poorer compared to the case when no correction was performed. They observed this phenomenon especially for models presenting large variance, and briefly suggested that this problem could be rooted to another bias source. It is indeed well-known in stochastic programming that the optimal values and solutions obtained with sample approximation are biased, due to the optimization operator.

It can be argued that bias often tends to disappear faster than variance. This bias can, however, becomes dominant is some applications. For instance, an interesting issue with simulated maximum likelihood is that the variance disappears when the number of observations (or individuals) grows to infinity, independently of the sample size used to approximate individual expectations, while the bias remains the same order as long as the number per observations does not change [2, 8, 18]. We will show here that the two biases (simulation and optimization) can counteract each other or have an additive effect, depending on the situation, and that care should be exercised due to this interaction. More generally, we set a more general framework of the simulation bias in optimization by relying on a stochastic programming formulation, and derive some observations about the interactions between the optimization and simulations biases.

The chapter is organized as follows. We present the general stochastic problem under consideration in Section 6.2, the bias estimator in Section 6.3, and its interaction with the optimization bias in Section 6.4, and illustrate it in Section 6.5. We conclude in Section 6.6.

6.2 Problem Formulation

We consider the general stochastic program

$$\min_{x \in \mathcal{X}} g\left(E[f(x,\xi)]\right), \tag{6.1}$$

where $\mathcal{X}$ is a compact set in $\mathcal{R}^n$, $\xi = (\xi_1, \dots, \xi_m)$ is a random vector of size m, $f : \mathcal{R}^n \times \mathcal{R}^m \to \mathcal{R}$, and $g : \mathcal{R} \to \mathcal{R}$. For simplicity, we assume that $\mathcal{X}$ does not depend on the random vector ξ, while it is easy to adapt the discussion to handle this case. We also denote $E[f(x,\xi)]$ by $f(x)$. In the stochastic programming literature, g is usually the identity function, so that the program (6.1) can be reduced to

$$\min_{x \in \mathcal{X}} E[f(x,\xi)].$$

Usually, one cannot compute exactly $E[f(x,\xi)]$, but has to approximate it, for instance by sampling over ξ. If the sample is kept the same during the entire optimization process, this leads to the heavily studied sample-path optimization technique, also know as the sample average approximation (SAA) or the stochastic counterpart method (see, e. g., [17, Chap. 4, and references therein]). In our context, with R random draws, one obtains the program

$$\min_{x \in \mathcal{X}} \hat{g}(x) = g\big(\hat{f}_R(x)\big) = g\left(\frac{1}{R}\sum_{r=1}^{R} f(x,\xi_r)\right). \tag{6.2}$$

When $g(x)$ is the identity, first-order consistency can be established under some appropriate assumptions. More precisely, the distance between the set of solutions of (6.2) and the set of solution of (6.1) almost surely converges to zero when the number of draws grows to infinity. Second-order still holds under the rigidity assumption, as discussed in [14], and shown in [3], along with some additional results. When $g(x)$ is not the identity (and is not affine), under the additional assumption that the range of f is compact, it is still possible to establish first-order consistency [4], as well as asymptotic normality. For finite R, however, g can produce a significant simulation bias, denoted by $B_R(x)$, and there is no guarantee of normality. In the rest of this chapter we will discuss a simple, yet often numerically cheap, way to estimate this bias and we will explore its use as a correction of the function to minimize. We conclude this section with two examples, the second being explored in more details in the numerical experiments.

Example 6.1. Consider the nonlinear programming problem

$$\min_{x} f(x)$$
$$\text{s.t. } E[c_i(x,\xi)] \geq 0,\ i = 1, \dots, s.$$

If one uses the log-barrier method in its simplest expression, one will replace this problem by a sequence of unconstrained problems of the form

$$\min_x f(x) - \mu \sum_{i=1}^{s} \ln E[c_i(x)],$$

which are solved for decreasing values of μ (see for instance Chap. 17 in Nocedal and Wright's textbook [13]).

Example 6.2. In discrete choice theory (see, among others, [5, 11]), a population of size I is associated with a set of J alternatives. Each individual i ($i = 1, \ldots, I$) has to choose one of these alternatives at one or multiple time periods, following the utility maximization principle: each alternative has a perceived utility and the individual selects the alternative with the highest utility. An external observer only has access to partial information and can only compute choice probabilities. The parameters are usually estimated by maximizing the corresponding likelihood:

$$\max_\theta L(\theta) = \prod_{i=1}^{I} p(i,\theta), \tag{6.3}$$

where θ is the vector of parameters to be estimated, and $p(i,\theta)$ reflects the probability of the choice or sequence of choices made by individual i. In order to reflect population heterogeneity, some coefficients in the utilities can be assumed to be randomly distributed over this population. Introducing the random vector $\boldsymbol{\xi}$, we rewrite (6.3) as

$$\max_\theta E\left[\prod_{i=1}^{I} p(i,\theta,\boldsymbol{\xi})\right].$$

The product $\prod_{i=1}^{I} p(i,\theta)$ is computationally unattractive, due to error propagations, but we can reformulate the problem in order to obtain better numerical properties. Consider first the form

$$\min_\theta -\frac{1}{I}\ln\left(E\left[\prod_{i=1}^{I} p(i,\theta;\boldsymbol{\xi})\right]\right),$$

or, assuming the individuals to be drawn independently,

$$\min_\theta -LL(\theta) = \frac{1}{I}\sum_{i=1}^{I} \ln\left(E\left[p(i,\theta;\boldsymbol{\xi}_i)\right]\right).$$

If R independent draws are used per individual, the SAA is

$$\min_\theta -\hat{LL}_R(\theta) = \frac{1}{I}\sum_{i=1}^{I} \ln\left(\hat{p}_R(i,\theta)\right) = \frac{1}{I}\sum_{i=1}^{I} \ln\left(\frac{1}{R}\sum_{r=1}^{R} p(i,\theta,\xi_{ir})\right). \tag{6.4}$$

A special case is the popular mixed logit model, defined by assuming that each observation follows a logit probability (see, e. g., [18, Chap. 6]).

6.3 Taylor-Based Bias Correction

It is tempting to estimate the bias introduced by g, when g is not the identity. This can be easily done when f and g are in C^2. Denoting by $h(x)$ the difference $\hat{f}_R(x) - f(x)$, we can develop the (statistical) Taylor expansion of $g(\hat{f}_R(x))$ around $g(f(x))$ as

$$g(\hat{f}_R(x)) = g(f(x)) + g'(f(x))h(x) + \frac{1}{2}g''(x)h^2(x) + O(h^3).$$

Since $E[h(x)] = 0$ and $E[h^2(x)] = \frac{1}{R}\text{Var} f(x, \boldsymbol{\xi})$, this implies

$$E\big[g\big(\hat{f}_R(x)\big)\big] - g(f(x)) = \frac{1}{2R}g''(f(x))\text{Var} f(x, \boldsymbol{\xi}) + O(E[h^3]),$$

so that an estimate of the bias can be computed as

$$\hat{B}_R(x) = \frac{1}{2R}g''(\hat{f}_R(x))\widehat{\text{Var}}[f(x, \boldsymbol{\xi})]. \tag{6.5}$$

Assuming that we can neglect the high-order terms, in particular that R is not too small, one can expect this estimate to be reliable, and construct the corrected problem

$$\min_{x \in \mathcal{X}} g\big(\hat{f}_R(x)\big) - \hat{B}_R(x), \tag{6.6}$$

especially when $g''(\cdot)$, and so $\hat{B}_R(x)$, can be computed at a reasonable cost. However, since $\hat{B}_R(x)$ is itself a statistical estimator, this can result in a variance increase, as

$$\begin{aligned}\text{Var}\big[g(\hat{f}_R(x)) - \hat{B}_R(x)\big] &= \text{Var}\big[g(\hat{f}_R(x))\big] + \text{Var}\big[\hat{B}_R(x)\big] \\ &\quad - \text{Cov}\big[(g\big(\hat{f}_R(x)\big), \hat{B}_R(x))\big].\end{aligned} \tag{6.7}$$

If $\text{Var}[\hat{B}_R(x)(x)]$ is high, and if the correlation between the bias and the sample average approximation is negative, this could result in a total variance increase. Moreover, this Taylor-based correction can itself be biased, making it difficult to quantify the variation of mean-square error (MSE) around a solution of the original or corrected problem. Therefore, it appears important to validate this approach for the application under consideration. We can proceed here with the bootstrap technique. Consider the sample $(\xi_1, \dots, \xi_R)$ used for the SAA, and construct the empirical distribution function $\hat{F}_R$ of $\boldsymbol{\xi}$. One samples R drawn with replacement from $\{\xi_1, \dots, \xi_R\}$, in order to obtain the new sample $\{\xi_1^b, \dots, \xi_R^b\}$, and calculates

$$\hat{B}_{R,b}(x) := \hat{B}_R\big(x, \xi_1^b, \dots, \xi_R^b\big).$$

Repeating the procedure q times at x^*, the solution of the optimization problem, one obtains the quantities $\hat{B}_{R,b_1}(x^*), \dots, \hat{B}_{R,b_q}(x^*)$. The variance of the bias estimator can be estimated as $\widehat{\text{Var}}[B_{R,b}]$, and its own bias as $E_{\hat{F}}[B_{R,b}(x^*)] - B_R(x^*)$. In practice, we use the improved bootstrap bias estimator in our numerical tests, and refer to [7] for a comprehensive coverage of these techniques.

From (6.7), one would prefer $\mathrm{Cov}[(g(\hat{f}_R(x)), \hat{B}_R(x))]$ to be positive. However, we assumed in the derivation of the bias estimator $\hat{B}_R(x)$ that we can chaply evaluate $g''(\hat{f}_R(x))$. This usually implies the reuse of elements obtained during the computation of the objective function and, implicitly, the use of common random numbers. If the induced correlation is positive, there is an effective variance reduction (but possibly only marginal). If the correlation is negative, one can try to use antithetic variates, using $1-U_r, r = 1, \ldots, R$ when computing $\hat{B}_R(x)$). This. however. makes the reuse of previously computed elements less direct, and sometimes not possible. The total computational cost therefore significantly increases, making the variance reduction ineffective compared to the simple procedure consisting in increasing the number of draws. The strategy to adopt is again application-dependent and can be determined by means of pilot experiments.

6.4 Impact on the Optimization Bias

Intuitively, it can be (and often is) assumed that solving (6.6) will produce more precise results than those obtained with the original problem. It is, however, well known in the stochastic programming literature that optimization produces a negative bias (see, e. g., [17, Chap. 4]):

$$E\big[\min_{x \in \mathcal{X}} \hat{f}_R(x)\big] \leq \min_{x \in \mathcal{X}} E\big[\hat{f}_R(x)\big] = f(x).$$

Similarly, using the definition of the simulation bias B_R, one has

$$E\big[\min_{x \in \mathcal{X}} g(\hat{f}_R(x))\big] \leq \min_{x \in \mathcal{X}} E\big[g(\hat{f}_R(x))\big] = \min_{x \in \mathcal{X}} [g(f(x)) + B_R(x)] \tag{6.8}$$

and

$$E\big[\min_{x \in \mathcal{X}} \big(g(\hat{f}_R(x)) - B_R(x)\big)\big] \leq \min_{x \in \mathcal{X}} E\big[\big(g(\hat{f}_R(x)) - B_R(x)\big)\big] = \min_{x \in \mathcal{X}} g\big(f(x)\big). \tag{6.9}$$

From these two expressions we can observe that the two bias sources interact, a phenomenon typically ignored in the estimation literature. Assume for the sake of discussion that $\hat{B}_R(x)$ is equal to $B_R(x)$, and that it keeps the same sign in the vicinity of the point of interest, typically a local solution of the corrected or uncorrected problem. We assume this solution to be global in $\mathcal{X}$, in order to ensure the inequalities to be valid. Not surprisingly, (6.9) indicates that if one could totally remove the simulation bias, one would still face an optimization bias. From (6.8) we can see that a negative simulation bias will amplify the optimization bias if not corrected. In this situation, the correction of the simulated bias appears very appealing. On the other hand, a positive simulation bias will play against the optimization bias, limiting the potential of the proposed correction, as these two biases could annihilate themselves if they have similar values.

It is, however, often very difficult to estimate the optimization bias and therefore judge the effect of the correction on the objective function and on the solution(s). Moreover, both biases change at different rates with the number of draws; the optimization bias initially originates from the variance of the objective function, while the simulation bias from the operator g. Increasing the number of draws typically reduces the bias contribution in the MSE faster than the variance, both of them being in $\mathcal{O}(1/R)$. This leads to a dominant optimization bias when the sample size is large enough, suggesting we should focus on variance reduction techniques when possible, rather than simulation bias. However, it can remain interesting to apply (6.5) as long as it is cheap to compute. Moreover, as illustrated below, some applications exist for which we can reduce the variance while not affecting the simulation bias, and correcting this bias can significantly improve the results.

6.5 Numerical Experiments

In this section, we will focus on mixed logit models estimation, as this problem motivated initial research on bias correction. Denote by ρ_i the relative error in $p(i, \theta)$:

$$\rho_i = \frac{\hat{p}_R(i,\theta) - p(i,\theta)}{p(i,\theta)},$$

It is easy to show (see, e. g., [12]) that

$$B_R(\theta) \approx \frac{1}{2I} \sum_{i=1}^{I} E[\rho_i^2] \tag{6.10}$$

$$\text{Var}\big[\hat{LL}_R(\theta)\big] \approx \frac{1}{I^2} \sum_{i=1}^{I} E[\rho_i^2]. \tag{6.11}$$

One has $E[\rho_i^2] = \mathcal{O}(1/R)$, so that the bias is $\mathcal{O}(1/R)$ and the variance is $\mathcal{O}(1/(RI))$. Thus, for fixed I, the contribution of the square bias to the MSE becomes negligible compared to that of the variance when R is large enough. But in practice, for computational budget issues, R is not always very large[1], so that the bias can be significant, and is not reduced when one increases I, in contrast to the variance. Moreover, many researchers currently advocate the use of panel data, that is when individuals deliver a sequence of choices rather than a unique choice. We empirically observed that the bias relative importance in the MSE was larger for such data, and that variance reduction techniques such as randomized quasi-Monte Carlo performed less efficiently [12]. For some realistic models, no significant improvement was even observed.

Observe from (6.10) and (6.11) that one can estimate the bias virtually for free, making the correction technique especially appealing in this case. We report some

[1] Many researchers even keep it to a fixed value, independently of the population size I.

Table 6.1. Log-likelihood values at the solution of the corrected maximization problem. The mean, standard deviation, and bootstrap bias are computed over 500 replications.

#*R*	500	500	1000	1000	2000	2000
Method	*standard*	*corrected*	*standard*	*corrected*	*standard*	*corrected*
Mean	−3.3173	−3.3060	−3.2968	−3.2905	−3.2886	−3.2849
Std dev	0.0079	0.0080	0.0061	0.0062	0.0046	0.0048
Bias	−0.0160	−0.0045	−0.0090	−0.0027	−0.0054	−0.0017

Table 6.2. Properties of bias estimator, evaluated at the solution of the corresponding maximum likelihood problem, computed over 500 bootstrap replications.

#*R*	500	500	1000	1000	2000	2000
Method	*standard*	*corrected*	*standard*	*corrected*	*standard*	*corrected*
Mean	−0.01080	−0.01142	−0.00616	−0.00632	−0.00372	−0.00372
Std dev	0.00049	0.00052	0.00036	0.00037	0.00032	0.00032
Bias	0.00097	0.00122	0.00066	0.00070	0.00058	0.00058

observations here obtained with a dataset of 4089 observations, delivered by 674 individuals, whose details are explained in [6]. The results are expressed with respect to the maximization of the log-likelihood, instead of a minimization program as in (6.4), in order to be consistent with the related literature. Table 6.1 illustrates the optimal values of $\hat{LL}_R(\theta)$, using the correction technique. In this table, "Std dev" stands for standard deviation. The uncorrected (standard) log-likelihood was also computed at the found solution, and the values compared. We can observe that the bias estimator indeed captures a significant part of the simulation bias, but still underestimates a bit its magnitude, as shown when computing the log-likelihood bias using a bootstrap procedure. The standard deviation of the log-likelihood function is barely affected by the correction, the bootstrap estimator having a very small variance, as indicated in Table 6.2.

The interaction with optimization can be observed in Figure 6.1, where we have restricted the number of observations to 1041 (delivered by 173 individuals) in order to face more variance, and so more optimization bias. If we only have to account for optimization bias, the value of the log-likelihood is decreasing for increasing numbers of draws, as expected, but when both biases are present, the behavior is not monotone, and the correction presents less benefit. It is sometimes advised by practitioners to increase the number of draws until the results stabilize in terms of objective function value. While this practice sounds reasonable, the previous figure also exhibits that the bias interaction can lead to an apparent early stabilization, and again, that care should be exercised in order to avoid false conclusions.

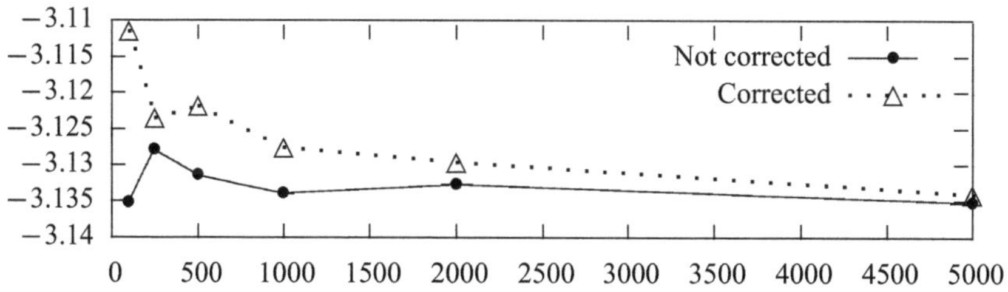

Figure 6.1. Evolution of the opposite log-likelihood optimal value, with and without the bias correction, averaged over 12 maximum likelihood replications, with a population of 173 individuals.

As stated above, we used common random numbers in order to save computational time, without consideration of the induced correlation. In order to judge if this approach was reasonable, we ran additional experiments and measured the correlation between the log-likelihood and the bias estimator. This correlation appeared to be negative, so that using antithetics variates could be more appropriate, as it would help to enforce a positive correlation. This, however, implies recomputing the probabilities instead of directly reusing the previous computations, meaning that the bias correction evaluation would cost approximately the same as the computation of the original objective function. In this example, the found covariance are too small to justify the additional cost. We nevertheless tried the approach, and obtained similar log-likelihood variances when estimating the bias using antithetics variates or common random numbers. We also conducted experiments with the real dataset used in [12] in order to confirm this observation, and obtained a covariance of $-1.9e^{-7}$ with common random variables, and of $5.28e^{-9}$ with antithetics variates, using 200 replications of the log-likelihood at the solution. This suggests only using antithetics variates to reduce the variance of the log-likelihood itself.

6.6 Conclusions

In thi chapter we have examined a class of nonlinear stochastic programming problems, characterized by an objective function defined as a nonlinear transformation of the expected value of another function. Since this problem is not attractive in its original form, we replace it by a sample average approximation, obtained by sampling over the underlying random vector. This problem still enjoys properties of consistency (under some regularity conditions), but typically suffers from a simulation bias. An estimator of this bias can be constructed using a second-order statistical Taylor development and computed at a cheap cost for some applications. We illustrated the method for the mixed logit estimation problem and reported experiments where the estimator indeed recovers most of the bias and does not introduce significant additional variance.

The main issue of such an approach, focused on bias correction rather than on variance reduction is that the simulation bias can interact with the optimization bias, possibly leading to a corrected optimal value with more bias than initially, a phenomenon not analysed previously. This can happen when these two types of bias are of opposite signs, so that for some range of sample sizes they tend to cancel out, while if they have the same sign, their effects are added, which fact pleads for the use of the proposed correction technique. The exact effect on the solutions rather than on the objective functions remains to be analyzed.

Acknowledgments. This research has been partly supported by FQRNT and NSERC grants.

References

[1] F. Bastin and C. Cirillo, Reducing simulation bias in mixed logit model estimation, *Journal of Choice Modelling* **3** (2010), 71–88.

[2] F. Bastin, C. Cirillo, and Ph, L. Toint, An adaptive Monte Carlo algorithm for computing mixed logit estimators, *Computational Management Science* **3** (2006), 55–79.

[3] F. Bastin, C. Cirillo, and Ph. L. Toint, Convergence theory for nonconvex stochastic programming with an application to mixed logit, *Mathematical Programming Series B* **108** (2006), 207–234.

[4] F. Bastin, C. Cirillo, and P. L. Toint, Formulation and solution strategies for nonparametric nonlinear stochastic programs, with an application in finance, *Optimization* **59** (2010), 355–376.

[5] M. Ben-Akiva and S. R. Lerman, *Discrete Choice Analysis: Theory and Application to Travel Demand*, MIT Press, Cambridge, MA, 1985.

[6] C. Cirillo and K. W. Axhausen, Evidence on the distribution of values of travel time savings from a six-week diary, *Transportation Research A* **40** (2006), 444–457.

[7] B. Efron and R. J. Tibshirani, *An Introduction to the Bootstrap*, Monographs on Statistics and Applied Probability 57, Chapman & Hall/CRC, Boca Raton, FL, 1993.

[8] C. Gouriéroux and A. Monfort, Simulation Based Econometrics in Models with Heterogeneity, *Annales d'Economie et de Statistiques* **20** (1991), 69–107.

[9] C. Gouriéroux and A. Monfort, *Simulation-based Econometric Methods*, Oxford University Press, Oxford, UK, 1996.

[10] D. Kristensen and B. Salanié, Higher Order Improvements for Approximate Estimators, Columbia University, Report no. 0910-15, Department of Economics, New York, NY, May 2010.

[11] D. L. McFadden, Modelling the Choice of Residential Location, in: A. Karlquist et al. (eds.), *Spatial Interaction Theory and Residential Location*, pp. 75–96, North Holland, Amsterdam, 1978.

[12] D. Munger, P. L'Ecuyer, F. Bastin, C. Cirillo, and B. Tuffin, Estimation of the Mixed Logit Likelihood Function by Randomized Quasi-Monte Carlo, *Transportation Research Part B*,**46** (2012), 305–320.

[13] J. Nocedal and S. J. Wright, *Numerical Optimization*, 2nd ed., Springer-Verlag, New York, 2006.

[14] R. Pasupathy, On choosing parameters in retrospective-approximation algorithms for stochastic root finding and simulation optimization, *Operations Research* **58** (2010), 889–901.

[15] S. M. Robinson, Analysis of sample-path optimization, *Mathematics of Operations Research* **21** (1996), 513–528.

[16] A. Shapiro, Monte Carlo Sampling Methods, in: A. Shapiro and A. Ruszczyński (eds.), *Stochastic Programming*, pp. 353–425, Handbooks in Operations Research and Management Science 10, Elsevier, 2003.

[17] A. Shapiro, D. Dentcheva, and A. Ruszczyński, *Lectures on Stochastic Programming*, SIAM, Philadelphia, PA, 2009.

[18] K. Train, *Discrete Choice Methods with Simulation*, Cambridge University Press, New York 2003.

[19] A. G. Tsagkanos, A bootstrap-based minimum bias maximum simulated likelihood estimator of mixed logit, *Economic Letters* **96** (2007), 282–286.

Author information

Fabian Bastin, University of Montreal, Montreal, Canada.
Email: `bastin@iro.umontreal.ca`

Chapter 7

Monte Carlo Simulation of Electron Transport in Quantum Cascade Lasers

Oskar Baumgartner, Zlatan Stanojević, and Hans Kosina

Abstract. A transport model for quantum cascade lasers based on the Pauli master equation is presented. An efficient Monte Carlo solver has been developed. The numerical methods to reduce the computational cost are discussed in detail. Finally, the simulator is used to obtain current-voltage characteristics as well as microscopic quantities of a mid-infrared QCL structure.

Keywords. Pauli Master Equation, Quantum Cascade Lasers, Electron Transport.

Mathematics Subject Classification 2010. 90C15, 90C30, 90C59, 62L20.

7.1 Introduction

Quantum cascade lasers (QCLs) offer a wide range of advantages which make them a popular choice for coherent light sources [5]. Their light emission is based on intersubband transitions. Due to the periodic nature of QCLs, a single electron will repeatedly contribute to the photon emission. The properties of the laser are mainly determined by the designer's choice of material and quantum well geometry.

For this purpose, simulation is a useful tool for tuning the QCL design to the desired optical and electrical characteristics. A requirement for such a simulator as design tool is a good balance between computational speed and physical accuracy. To describe the electronic properties of the laser, a quantum mechanical transport model is necessary. Previously the nonequilibrium Green's function formalism (NEGF) has been used as a rigorous approach to capture the QCL's physics [10, 11]. Unfortunately the inherently high computational costs of the NEGF formalism render it unfeasible as a design tool.

7.2 QCL Transport Model

In our approach we use the Pauli master equation (PME) [4] to model current transport through the QCL's semiconductor heterostructure. Based on the experiences of a MATLAB prototype presented in [14], an optimized Monte Carlo (MC) simulator has been implemented in C++ within the Vienna–Schrödinger–Poisson (VSP) simulation framework [9].

7.2.1 Pauli Master Equation

Theoretical studies showed that in many practical cases the steady state transport in QCLs is incoherent such that a semiclassical description can be employed [7, 8]. Following this approach, we developed a transport simulator for quantum cascade lasers based on the Pauli master equation [14]. The transport is described via scattering transitions among quasistationary basis states which are determined by numerically solving the Schrödinger equation. The Hamiltonian includes the band edge formed by the heterostructure. In this way, tunneling is accounted for through the delocalized eigenstates.

The transport equations are derived from the Liouville–von Neumann equation in the Markov limit in combination with the diagonal approximation. This means that the off-diagonal elements of the density matrix are neglected and one arrives at the Boltzmann-like Pauli master equation [6]

$$\frac{\mathrm{d} f_{\mathbf{k},n}(t)}{\mathrm{d}t} = \sum_{\mathbf{k}',m} \left\{ S_m^n\left(\mathbf{k}',\mathbf{k}\right) f_{\mathbf{k}',m}(t) - S_n^m\left(\mathbf{k},\mathbf{k}'\right) f_{\mathbf{k},n}(t) \right\}.$$

Here, m and n denote the subband indices, and $\mathbf{k}$ and $\mathbf{k}'$ the in-plane wave vectors. The transition rate from state $|\mathbf{k}', m\rangle$ to state $|\mathbf{k}, n\rangle$ for an interaction H_{int} follows from Fermi's golden rule

$$S_n^m\left(\mathbf{k},\mathbf{k}'\right) = \frac{2\pi}{\hbar} \left|\langle \mathbf{k}', m | H_{\text{int}} | \mathbf{k}, n\rangle\right|^2 \delta\left(\mathcal{E}(\mathbf{k}') - \mathcal{E}(\mathbf{k}) \mp \hbar\omega\right).$$

We make use of the translational invariancy of the QCL structure and simulate the electron transport over a single stage only. The wave function overlap between the central stage and spatially remote stages is small. It is therefore assumed that interstage scattering is limited only to the nearest neighbor stages and that interactions between basis states of remote stages can be safely neglected.

The states of the whole QCL device structure are assumed to be a periodic repetition of the states of a central stage. This approach ensures charge conservation and allows imposing periodic boundary conditions on the Pauli master equation.

Since transport is simulated over a central stage only, every time a carrier undergoes an interstage scattering process the electron is reinjected into the central stage with an energy changed by the voltage drop over a single period. The total current is determined by the net number of interstage transitions.

The transport equations can be solved using a Monte Carlo approach. We developed an algorithm and devised several new numerical methods to reduce the computational cost of the simulation. The implementation details will be discussed in Section 7.2.3.

7.2.2 Calculation of Basis States

The task at hand can be divided into two parts. First, the basis states need to be determined. Second, the states have to be assigned to a stage according to their periodicity. For this purpose the equation definition and solver facilities of the VSP were used beneficially.

Since it is essential to consider band nonparabolicity for QCLs, the user can choose one of several models for the Hamiltonian of the Schrödinger equation. For this purpose, in addition to the single band effective mass model, an effective two-band **k·p** model [16] or a three-band **k·p** model are available.

To describe the openness of the quantum system we make use of the perfectly matched layer (PML) boundary conditions for the Schrödinger equation [15]. Perfectly matched layers were originally used as boundary conditions for electromagnetic and waveguide problems [1]. The PML boundary conditions give rise to a complex eigenvalue problem. It is solved by means of Arnoldi iteration and the ARPACK [12] library linked to VSP. The calculated eigenvectors correspond to the complex wavefunctions. The real part of the eigenvalue is the eigenenergy of the quasibound state, whereas the imaginary part is related to its finite lifetime due to the openness of the system. This allows us to estimate the tunneling current using the following relation [2]:

$$J_{\text{Tunnel}} = \sum_i \frac{n_i}{\tau_i}.$$

As long as the hereby calculated tunneling current is small compared to the incoherent current, the semiclassical approach of the PME remains valid.

We calculated the eigenstates for an $In_{0.53}Ga_{0.47}As$ / $GaAs_{0.51}Sb_{0.49}$ mid infrared (MIR) quantum cascade laser reported in [3]. The barrier thicknesses (in bold) and the well thicknesses of one period in nanometer are **7.5**/2.9/**1.5**/6.5/ **1.8**/5.9/**5.5**/5.6/**2.6**/ 5.1/**2.0**/<u>5.2</u>/**4.0**/4.2/**2.9**/4.4. The underlined layer is doped to a concentration of $4 \times 10^{17}\text{cm}^{-3}$. We will use this device as a benchmark throughout this chapter. The calculated wavefunctions for a two-band **k·p** Hamiltonian with PML boundary conditions are shown in Figure 7.1.

To use the eigenvectors as basis states in the MC routine we need to consider the periodicity of the device and automatically select the field-periodic states of a single stage. For that purpose we calculate the crosscorrelation and autocorrelation $C_{ij}(x) = \Psi_i(x) \star \Psi_j(x)$ of all subbands. We make use of the relation $\mathcal{F}\{\Psi_i \star \Psi_j\} = \mathcal{F}\{\Psi_i\}^* \cdot \mathcal{F}\{\Psi_j\}$ and the Fast Fourier Transform to quickly obtain the correlations, i. e., $C_{ij}(x) = \mathcal{F}^{-1}\{\mathcal{F}\{\Psi_i\}^* \cdot \mathcal{F}\{\Psi_j\}\}$. Then the positions $x_{ij,\max}$ of the maxima of the correlations $C_{ij}(x)$ are determined for all the subbands. If $x_{ij,\max}$ lies at the geometric period length L of the QCL structure, the two states i and j are considered field-periodic and given an appropriate stage index. As an example the field-periodic states of the InGaAs/GaAsSb QCL are given in Figure 7.2.

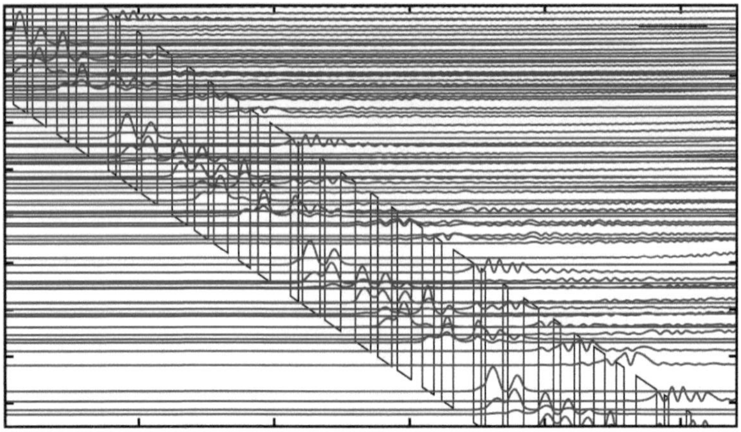

Figure 7.1. Multiple cascades of a QCL need to be considered om order to obtain suitable basis states for the PME Monte Carlo solver.

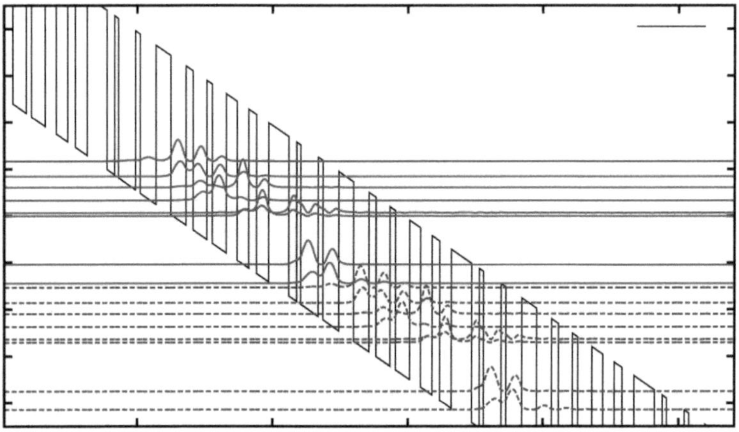

Figure 7.2. Application of our subband selection routine which automatically assigns the field-periodic wavefunctions to a stage of the QCL (only the states of two stages are shown).

7.2.3 Monte Carlo Solver

The wavefunctions provided by the routines discussed above are processed to initialize the MC code. After calculating the scattering rates using functor classes and filling the data structures, the initial valley, subband, and energy of the carrier are randomly selected. In the MC loop the precalculated possible scattering processes are looked up for the actual electron state. A random number r is determined using a uniform distribution in the interval $[0, P_n]$ where $P_n = \sum_{j=1}^{n} \Gamma_j$ is the total scattering rate. The scattering process i is selected from the table such that the relation $P_{i-1} < r \leq P_i$

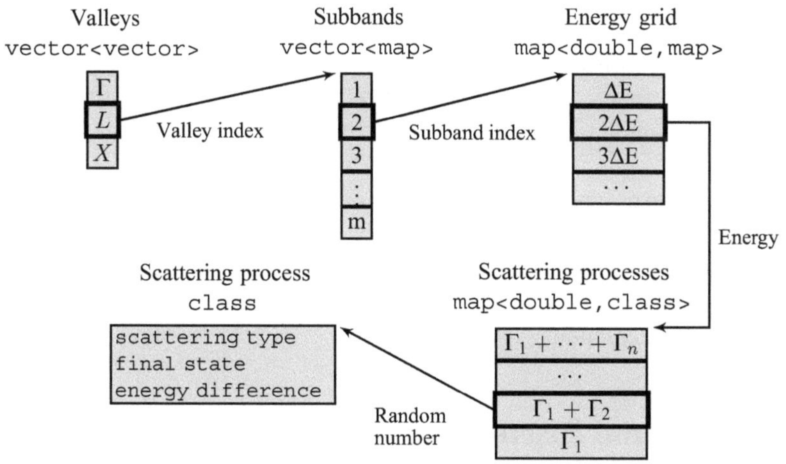

Figure 7.3. Data structures for the selection of a scattering process. Valleys and subbands are accessed by index. Each subband uses its own energy grid. For a fast lookup the grid is implemented as a C++ standard template library (STL) `map` with the energy as key value. Similarly, scattering processes are stored in a map where the partial sums of their transition rates are used as key. The selected scattering process instance contains all essential information to update the statistics and the state variable.

holds, where the P_i are the partial sums of the scattering rates. The data structure for the selection method is given in Figure 7.3. As shown, the C++ standard template library containers are used with regard to minimizing the lookup time.

The actual state and the chosen scattering process are used to update the statistical quantities such as subband population, energy distribution, and current. Afterwards the state variable is set to its new value given by the scattering event. The MC loop is terminated when the given number of events is reached.

To account for the periodic structure of the device, the subbands of three stages are included. Whenever the electron scatters from the central to the left or the right stage it is reinjected into the corresponding state of the central stage, and the estimator for the electrical current is updated.

We identified the calculation of the polar-optical phonon scattering rate as one of the major contributions to the simulation run time. Therefore, we optimized the calculation of the scattering rate for this process by exchanging the order of the multiple integrations. An analytic integration over the final states is carried out first. The integration related to the matrix element is carried out last. The remaining integration is in momentum space and has the form

$$\Gamma_{mn}(\mathbf{k}_\parallel) = \frac{m^*}{\hbar^2}\frac{e^2\omega_{\mathrm{PO}}}{4\pi\varepsilon}\left(n_{\mathrm{PO}} + \frac{1}{2} \mp \frac{1}{2}\right)\int \frac{|\widehat{\rho}_{mn}(q_z)|^2}{\sqrt{(k_\parallel^2 + k_\mathrm{f}^2 + q_z^2)^2 - 4k_\parallel^2 k_\mathrm{f}^2}}\mathrm{d}q_z,$$

where $k_f^2 = k_{\|}^2 + \frac{2m^*}{\hbar^2}(\mathcal{E}_m - \mathcal{E}_n \pm \hbar\omega_{PO})$ has to be positive in order to satisfy energy conservation. This allows us to use a Fast Fourier Transform (FFT) to calculate the overlap integrals $\widehat{\rho}_{mn}(q_z) = \mathcal{F}\{\rho_{mn}(z)\}$, where $\rho_{mn}(z) = \psi_m^*(z)\psi_n(z)$, which reduces the calculation time of the PO scattering rate by three to four orders of magnitude.

Currently, acoustic and optical deformation potential and polar optical electron-phonon scattering as well as alloy, intervalley, and interface roughness scattering are included. The object-oriented implementation allows for simple inclusion of additional physics for further investigation of QCL devices such as electron–electron and electron–photon interaction.

7.3 Results and Discussion

We used the implemented transport model to simulate the InGaAs/GaAsSb mid infrared (MIR) quantum cascade laser reported in [3]. The current density as a function of the electric field at 90 Kelvin is shown in Figure 7.4. The simulation result is in reasonable agreement with the experiment. The maximum in the current around the laser's designed optimum field strength can be attributed to a PO phonon resonance of the lower and ground laser level. The resonance causes a fast depletion of the lower laser level and aids the population inversion. The decrease in current above the threshold is due to increasing coherent tunneling to the continuum as well as the electron-photon interaction, which is not yet included in the model. At low fields the current is

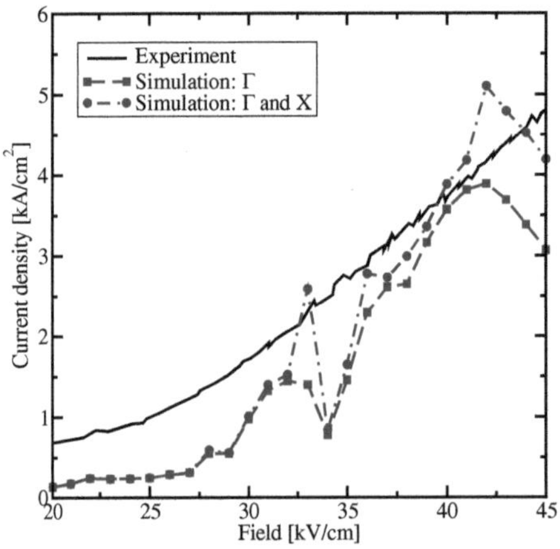

Figure 7.4. Current density vs. applied electric field of the MIR QCL. The simulation shows that the inclusion of the X valley has only a small influence on the characteristics around the laser threshold.

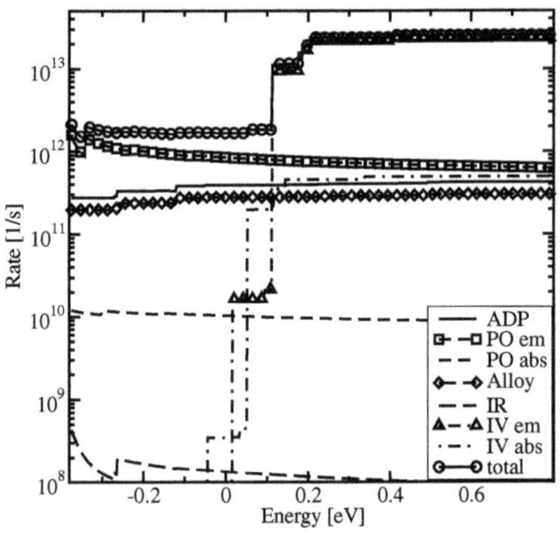

Figure 7.5. Calculated scattering rates with the lower laser level as initial subband. At energies near the subband minimum the polar optical emission is the dominant process. This ensures fast depopulation of the lower laser level.

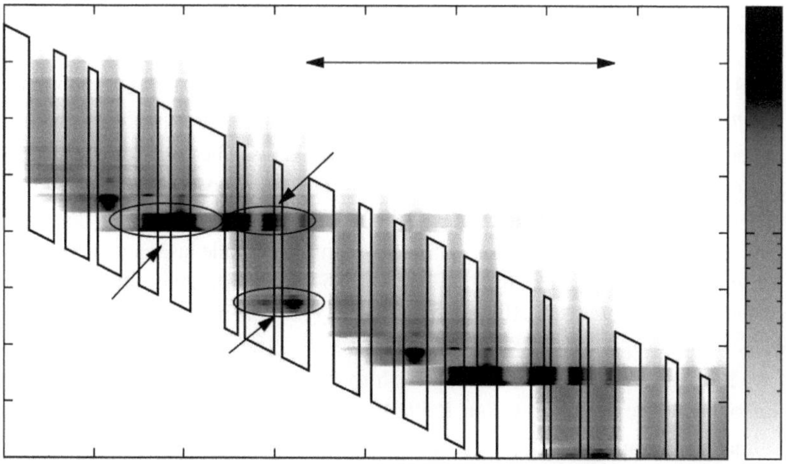

Figure 7.6. Conduction band edge and carrier density spectrum obtained by the Pauli master equation solver at an electric field strength of 40 kV/cm. The occupation of the upper laser state is clearly visible.

underestimated due to the dominant coherent transport through the barriers [13]. The characteristics show that transport in the X valley contributes only marginally to the total current near the laser threshold. This is also indicated by the scattering rates for the lower laser level (Figure 7.5), where PO emission is also shown to be dominant.

The calculation of a single operating point typically takes a few minutes, depending on number of valleys, subbands, and energy grid resolution. This is orders of magnitude faster than a full quantum treatment using nonequilibrium Green's functions, but still gives insight to microscopic quantities such as the carrier density spectrum shown in Figure 7.6.

7.4 Conclusion

We have presented a semiclassical transport model for quantum cascade lasers based on the Pauli master equation. We devised new numerical methods to reduce the computational demand and realized an efficient Monte Carlo simulator implemented in C++. The model was applied to a mid infrared QCL. It gives insight to macro- and microscopic quantities such as current-voltage characteristics, scattering rates, carrier density spectrum, subband population, and optical gain.

Acknowledgments. This work was supported by the Austrian Science Fund special research program IR-ON (F2509).

References

[1] J.-P. Berenger, A perfectly matched layer for the absorption of electromagnetic waves, *J. Comput. Phys.* **114** (1994), 185–200.

[2] R. Clerc, A. Spinelli, G. Ghibaudo, and G. Pananakakis, Theory of direct tunneling current in metal–oxide–semiconductor structures, *J. Appl. Phys.* **91** (2002), 1400–1409.

[3] H. Detz, M. Nobile, C. Deutsch, P. Klang, A. Andrews, W. Schrenk, K. Unterrainer, and G. Strasser, Improved InGaAs/GaAsSb quantum cascade laser active region designs, *Journal of Modern Optics* **58**(21) (2011), 2015–2020, available online at http://www.tandfonline.com/doi/abs/10.1080/09500340.2011.604734 (last accessed September 21, 2012).

[4] M. V. Fischetti, Master-equation approach to the study of electronic transport in small semiconductor devices, *Phys. Rev. B* **59** (1999), 4901–4917.

[5] C. Gmachl, F. Capasso, D. L. Sivco, and A.Y. Cho, Recent progress in quantum cascade lasers and applications, *Rep. Progr. Phys.* **64** (2001), 1533–1601.

[6] R. C. Iotti, E. Ciancio, and F. Rossi, Quantum transport theory for semiconductor nanostructures: A density-matrix formulation, *Phys. Rev. B* **72** (2005), 125347.

[7] R. C. Iotti and F. Rossi, Nature of Charge Transport in Quantum-Cascade Lasers, *Phys. Rev. Lett.* **87** (2001), 146603.

[8] C. Jirauschek, G. Scarpa, P. Lugli, M. S. Vitiello, and G. Scamarcio, Comparative analysis of resonant phonon THz quantum cascade lasers, *J. Appl. Phys.* **101** (2007), 086109.

[9] M. Karner, A. Gehring, S. Holzer, M. Pourfath, M. Wagner, W. Goes, M. Vasicek, O. Baumgartner, C. Kernstock, K. Schnass, G. Zeiler, T. Grasser, H. Kosina, and S. Selberherr, A Multi-Purpose Schrödinger-Poisson Solver for TCAD Applications, *J. Comput. Electron.* **6** (2007), 179–182.

[10] T. Kubis and P. Vogl, Self-consistent quantum transport theory: Applications and assessment of approximate models, *J. Comput. Electron.* **6** (2007), 183–186.

[11] S. C. Lee and A. Wacker, Quantum transport calculations for quantum cascade laser structures, *Physica E* **13** (2002), 858–861.

[12] R. B. Lehoucq, D. C. Sorensen, and C. Yang, *ARPACK Users' Guide: Solution of Large-Scale Eigenvalue Problems with Implicitly Restarted Arnoldi Methods*, Society for Industrial and Applied Mathematics, 1998.

[13] A. Mátyás, T. Kubis, P. Lugli, and C. Jirauschek, Comparison between semiclassical and full quantum transport analysis of THz quantum cascade lasers, *Physica E: Low-dimensional Systems and Nanostructures* **42** (2010), 2628–2631, 14th International Conference on Modulated Semiconductor Structures.

[14] G. Milovanovic and H. Kosina, A semiclassical transport model for quantum cascade lasers based on the Pauli master equation, *J. Comput. Electron.* **9** (2010), 211–217.

[15] S. Odermatt, M. Luisier, and B. Witzigmann, Bandstructure calculation using the **k**·**p** method for arbitrary potentials with open boundary conditions, *J. Appl. Phys.* **97** (2005), 046104.

[16] C. Sirtori, F. Capasso, J. Faist, and S. Scandolo, Nonparabolicity and a sum rule associated with bound-to-bound and bound-to-continuum intersubband transitions in quantum wells, *Phys. Rev. B* **50** (1994), 8663–8674.

Author information

Oskar Baumgartner, Institute for Microelectronics, Vienna University of Technology, Vienna, Austria.
Email: baumgartner@iue.tuwien.ac.at

Zlatan Stanojević, Institute for Microelectronics, Vienna University of Technology, Vienna, Austria.
Email: stanojevic@iue.tuwien.ac.at

Hans Kosina, Institute for Microelectronics, Vienna University of Technology, Vienna, Austria.
Email: kosina@iue.tuwien.ac.at

Chapter 8

Markov Chain Monte Carlo Particle Algorithms for Discrete-Time Nonlinear Filtering

Avishy Carmi and Lyudmila Mihaylova

Abstract. This work shows how a carefully designed instrumental distribution can improve the performance of a Markov chain Monte Carlo (MCMC) particle filter for systems with a high state dimension (up to 100). We devise a special subgradient-based kernel from which candidate moves are drawn. This facilitates the implementation of the filtering algorithm in high-dimensional settings using a remarkably small number of particles. We demonstrate our approach in solving a nonlinear non-Gaussian dynamic compressed sensing (l_1 constrained) problem and show high estimation accuracy.

Keywords. Markov Chain Monte Carlo Methods, Nonlinear Filtering, High Dimensional Systems.

Mathematics Subject Classification 2010. 62Lxx, 62Mxx, 93E10, 60G35, 60J05, 60J20, 90B50, 62Gxx, 65C05, 93Exx.

8.1 Introduction

Over the past decade we have witnessed an immense rise in the complexity of inference tasks, primarily owing to recent advancements in computation power and ever increasing performance expectations. Contemporary filtering algorithms nowadays confront the curse of dimensionality in many real world applications which include, among others, multiple object and crowd tracking [2, 4], learning algorithms, and reasoning in multiagent systems. In many typical problems in these and similar domains, even the most conservative restrictions on the underlying spatiotemporal complexities would not alleviate the dimensionality issue.

A thoughtful study of the class of sequential Monte Carlo methods, otherwise known as particle filters (PFs), has shown that it is the well-known importance sampling technique which renders most state-of-the-art PF schemes inadequate for reasoning in complex high dimensional settings [8]. Recently, various approaches have been proposed for overcoming this downfall. These account for MCMC methods [5, 6], population PFs and log-homotopy particle flow [9].

In this chapter we derive an efficient version of a genuine and simple MCMC particle filtering algorithm, several variants of which appeared in the literature. In particular, our proposed methodology endows the plain MCMC filtering algorithm with an informative instrumental density from which new moves are generated. As part of this, the local subgradient of the likelihood is exploited for potentially steering the produced chain to highly probable regions of the exploration space. This approach possesses an advantage over the prevalent particle refinement technique which utilizes a Metropolis-within-Gibbs stage for making conditional draws – a stage which has been conjectured to deteriorate the chain mixing time [5]. We demonstrate the potential of our approach in complex settings involving nonlinear state dynamics. The rest of the chapter is organised in the following way: Section 8.2 presents the problem formulation, Section 8.3 describes MCMC filtering techniques for high dimensional systems, Section 8.4 gives illustrative examples, and finally Section 8.5 summarizes the results.

8.2 General Particle Filtering Framework

Consider the problem of estimating the state of a dynamic system:

$$\mathbf{x}_k = f(\mathbf{x}_{k-1}, \mathbf{v}_{k-1}), \tag{8.1a}$$

$$\mathbf{z}_k = h(\mathbf{x}_k, \mathbf{r}_k), \tag{8.1b}$$

where $\mathbf{x}_k$ is the unknown system state vector, and $f(.)$ and $h(.)$ are the process and measurement functions, respectively. The above equations are driven by $\mathbf{v}_k$ and $\mathbf{r}_k$ which stand for the process and observation noises.

According to Bayes' rule, the filtering probability density function (pdf) $p(\mathbf{x}_k|\mathbf{z}_{1:k})$ of the state vector $\mathbf{x}_k \in \mathbb{R}^{n_x}$ given the measurement history $\mathbf{z}_{1:k} = \{\mathbf{z}_1, \dots, \mathbf{z}_k\}$ may be written as

$$p(\mathbf{x}_k|\mathbf{z}_{1:k}) = \frac{p(\mathbf{z}_k|\mathbf{x}_k)p(\mathbf{x}_k|\mathbf{z}_{1:k-1})}{p(\mathbf{z}_k|\mathbf{z}_{1:k-1})}, \tag{8.2}$$

where $p(\mathbf{z}_k|\mathbf{z}_{1:k-1})$ is the normalising constant. The state predictive distribution is given by the Chapman–Kolmogorov equation

$$p(\mathbf{x}_k|\mathbf{z}_{1:k-1}) = \int_{\mathbb{R}^{n_x}} p(\mathbf{x}_k|\mathbf{x}_{k-1})p(\mathbf{x}_{k-1}|\mathbf{z}_{1:k-1})d\mathbf{x}_k. \tag{8.3}$$

The evaluation of the right hand side of (8.2) involves integration which can be avoided in the particle filtering approach by approximating the filtering pdf $p(\mathbf{x}_k|\mathbf{z}_{1:k})$ with a set of particles $\mathbf{x}_{0:k}^{(i)}$, $i = 1, \dots, N$ and their corresponding weights $w_k^{(i)}$ [1]. Then the posterior density can be written as follows:

$$p(\mathbf{x}_{0:k}|\mathbf{z}_{1:k}) = \sum_{i=1}^{N} w_k^{(i)} \delta(\mathbf{x}_{0:k} - \mathbf{x}_{0:k}^{(i)}), \tag{8.4}$$

where $\delta(.)$ is the Dirac delta function, and the weights are normalised such that $\sum_i w_k^{(i)} = 1$. Each pair $\{\mathbf{x}_{0:k}^{(i)}, w_k^{(i)}\}$ characterizes the belief that the system is in state

$\mathbf{x}_{0:k}^{(i)}$. An estimate of the variable of interest is obtained by the weighted sum of particles. Two major stages can be distinguished: *prediction* and *update*. During prediction, each particle is modified according to the state model, including the addition of random noise in order to simulate the effect of the noise on the state. In the update stage, each particle's weight is reevaluated based on the new data. A *resampling* procedure introduces variety in the particles by eliminating those with small weights and replicating the particles with larger weights such that the approximation in (8.4) still holds. The residual resampling algorithm [3] is applied here. This is a two step procedure which makes use of the sampling-importance-resampling (SIR) scheme. This chapter proposes a resample-move scheme which is presented in detail as Algorithms 8.1 and 8.2.

8.3 High Dimensional Particle Schemes

The importance sampling approach, which essentially forms the core of every PF algorithm, becomes prohibitively inefficient in high dimensions [8]. Over the past decade this caveat has motivated the derivation of far more sophisticated particle schemes, most of which rely on MCMC techniques [4–6]. The main purpose of this paper is to corroborate and extend a single promising direction in this regard. We demonstrate the strong potential of a class of genuine MCMC-based particle algorithms.

8.3.1 Sequential MCMC Filtering

The following sequential filtering scheme is closely related to the inference algorithms presented in [2, 4] (see also [6]). Suppose that at time $k-1$ there are N samples $\{\mathbf{x}_{k-1}^{(i)}\}_{i=1}^{N}$ drawn approximately from the filtering density $p(\mathbf{x}_{k-1} \mid \mathbf{z}_{1:k-1})$ (i. e., the previous time target distribution). A new set of samples $\{\mathbf{x}_{k}^{(i)}\}_{i=1}^{N}$ representing $p(\mathbf{x}_k \mid \mathbf{z}_{1:k})$ can be then simulated using a tailored Metropolis Hastings (MH) scheme.

The MH algorithm generates samples from an aperiodic and irreducible Markov chain with a predetermined (possibly unnormalised) stationary distribution. This is a constructive method which specifies the Markov transition kernel by means of acceptance probabilities based on the preceding time outcome. Setting the stationary density as the joint filtering pdf $p(\mathbf{x}_k, \mathbf{x}_{k-1} \mid \mathbf{z}_{1:k})$ (the marginal of which is the desired filtering pdf $p(\mathbf{x}_k \mid \mathbf{z}_{1:k})$), a new set of samples from this distribution can be obtained after the MH burn-in period. This procedure is described next.

First we simulate a sample $\mathbf{x}'_k$ from the joint pdf $p(\mathbf{x}_k, \mathbf{x}_{k-1} \mid \mathbf{z}_{1:k-1})$ by drawing

$$\mathbf{x}'_k \sim p(\mathbf{x}_k \mid \mathbf{x}'_{k-1}), \tag{8.5}$$

where $\mathbf{x}'_{k-1}$ is uniformly drawn from the empirical approximation of $p(\mathbf{x}_{k-1} \mid \mathbf{z}_{1:k-1})$ given by

$$\hat{p}(\mathbf{x}_{k-1} \mid \mathbf{z}_{1:k-1}) = N^{-1} \sum_{i=1}^{N} \delta(\mathbf{x}_{k-1}^{(i)} - \mathbf{x}_{k-1}). \tag{8.6}$$

This sample $\mathbf{x}'_k$ is accepted or rejected using the following Metropolis rule.

Let $(\mathbf{x}_k^{(i)}, \mathbf{x}_{k-1}^{(i)})$ be a sample from the realised chain of which the stationary distribution is the joint filtering pdf. The MH algorithm accepts the new candidate pair $(\mathbf{x}'_k, \mathbf{x}'_{k-1})$ as the next realisation from the chain with probability

$$\alpha = \min\left\{1, p(\mathbf{z}_k \mid \mathbf{x}'_k)/p(\mathbf{z}_k \mid \mathbf{x}_k^{(i)})\right\}, \tag{8.7}$$

i. e.,

$$(\mathbf{x}_k^{(i+1)}, \mathbf{x}_{k-1}^{(i+1)}) = \begin{cases} (\mathbf{x}'_k, \mathbf{x}'_{k-1}), & \text{if } u \leq \alpha \\ (\mathbf{x}_k^{(i)}, \mathbf{x}_{k-1}^{(i)}), & \text{otherwise,} \end{cases} \tag{8.8}$$

with the uniform random variable $u \sim U[0, 1]$. The above sampling scheme may be inefficient in exploring the sample space, since the underlying proposal density of a well-behaved system (i. e., of which the process noise is of low intensity) introduces relatively small moves. This drawback can be alleviated by a secondary Metropolis-within-Gibbs refinement stage [2, 4].

8.3.2 Efficient Sampling in High Dimensions

The efficiency of MCMC samplers, which mainly refers to the mixing properties of the produced chain, is prominently affected by the proposal density from which the candidate moves are drawn. Good proposals facilitate the exploration of the sample space and in particular of high probability regions irrespectively of the initial conditions. The mixing problem is partially alleviated in both [4] and [2], where Metropolis within Gibbs refinement stages are incorporated into the basic MCMC scheme. This approach has proved itself viable for various multiobject tracking applications.

In this chapter we investigate a rather different type of proposal in which the (sub) gradient information of the likelihood is taken into account. As demonstrated in the following, this unique proposal facilitates the application of MCMC filtering in high dimensional state spaces (as far as particle filtering is concerned) using a remarkably small number of particles. The idea consists of constructing a proposal out of a set of improved samples using the joint propagated pdf $p(\mathbf{x}_k, \mathbf{x}_{k-1} \mid \mathbf{z}_{1:k-1})$. Thus, the obtained samples from (8.5) are pushed towards high probability regions based on the (sub)gradient of the likelihood

$$\bar{\mathbf{x}}_k^{(i)} = \mathbf{x}_k'^{(i)} - \lambda^{(i)} \frac{\log p\left(\mathbf{z}_k \mid \mathbf{x}_k'^{(i)}\right)}{\| t^{(i)} \|_2^2} t^{(i)}, \quad i = 1, \ldots, N, \tag{8.9}$$

where $\|\ .\ \|$ denotes Euclidean norm, the relaxation parameter $\lambda^{(i)} \sim p_\lambda$ is sampled for every i from some prescribed *steering distribution* p_λ (e. g., uniform), and $t^{(i)} := \partial \log p(\mathbf{z}_k \mid \mathbf{x}_k)$ is the associated subgradient with respect to $\mathbf{x}_k$, computed at $\mathbf{x}'^{(i)}_k$. This technique is essentially related to the acclaimed iterative convex optimisation method known as subgradient projection [10]. Having the set of improved and propagated particles, $\{\bar{\mathbf{x}}^{(i)}_k, \mathbf{x}'^{(i)}_{k-1}\}_{i=1}^N$, a regularized proposal is constructed in the following manner:

$$q(\bar{\mathbf{y}}_k) \propto \sum_{i=1}^{N} \mathcal{N}(\bar{\mathbf{y}}_k \mid \bar{\mathbf{y}}^{(i)}_k, \sigma^2),\ \ \bar{\mathbf{y}}_k = [\bar{\mathbf{x}}^T_k, \mathbf{x}'^T_{k-1}]^T,\ \bar{\mathbf{y}}^{(i)}_k = [(\bar{\mathbf{x}}^{(i)}_k)^T, (\mathbf{x}'^{(i)}_{k-1})^T]^T, \tag{8.10}$$

where $\mathcal{N}(\cdot)$ and σ denote the normal distribution and a roughening intensity parameter, respectively. An MH procedure is then carried out in a fashion similar to (8.8). This time, however, the acceptance probability of a new candidate pair $(\bar{\mathbf{x}}_k, \mathbf{x}'_{k-1}) \sim q(\bar{\mathbf{y}}_k)$, is given by

$$\alpha = \min\left\{1, \frac{p(\mathbf{z}_k \mid \bar{\mathbf{x}}_k)\hat{p}(\bar{\mathbf{x}}_k, \mathbf{x}'_{k-1} \mid \mathbf{z}_{1:k-1})q(\mathbf{y}^{(i)}_k)}{p(\mathbf{z}_k \mid \mathbf{x}^{(i)}_k)\hat{p}(\mathbf{x}^{(i)}_k, \mathbf{x}^{(i)}_{k-1} \mid \mathbf{z}_{1:k-1})q(\bar{\mathbf{y}}_k)}\right\}, \tag{8.11}$$

where $\bar{\mathbf{y}}_k) = [\mathbf{x}^T_k, \mathbf{x}^T_{k-1}]^T$ is a vector containing the pair before the mix.

8.3.3 Setting Proposal and Steering Distributions

In the above scheme the proposal distribution $q(\bar{\mathbf{y}}_k)$ is obtained prior to the application of the MH stage. This computationally excessive, though necessary, step is used in conjunction with the MH for producing a reversible chain that will ultimately converge to the prescribed stationary distribution. An efficient alternative for computing $q(\bar{\mathbf{y}}_k)$ is by replacing (8.10) with a single Gaussian of which the statistical moments correspond to the sample mean and covariance of the population $\{\bar{\mathbf{y}}^{(i)}_k\}_{i=1}^N$. A pseudocode of this variant of the MCMC particle filter is provided in Algorithm 8.1.

An approach which is likely to have an improved MH acceptance rate relies on using two distinct proposals $q_l(\bar{\mathbf{y}}_k)$, $l = 1, 2$, each of which involves a different steering distribution $p^l_\lambda(\lambda)$. The new moves are then sampled from each of these two proposals in an alternating fashion. This approach, which is provided here without a theoretical justification, is summarized in Algorithm 8.2.

8.4 Illustrative Examples

In the following examples we compare the performance of a few nonlinear filtering algorithms applied to systems with $n_x = 100$ states. The filters refer to an extended Kalman filter (EKF), the compressed sensing Kalman filter (CSKF) of [7], and the proposed MCMC particle filtering algorithm (we have left out the performance results

Algorithm 8.1. MCMC Particle-filtering algorithm.

1: Given previous time samples $\mathbf{x}_{k-1}^{\prime(i)}$, $i = 1, \dots, N$ perform the following steps.
2: Draw $\mathbf{x}_k^{\prime\,(i)} \sim p(\mathbf{x}_k \mid \mathbf{x}_{k-1}^{\prime(i)})$, $i = 1, \dots, N$.
3: Use (8.9) for producing $\bar{\mathbf{x}}_k^{(i)}$, $i = 1, \dots, N$. The set $\{\mathbf{x}_k^{\prime\,(i)}, \mathbf{x}_{k-1}^{\prime(i)}\}_{i=1}^N$ simulates $\hat{p}(\mathbf{x}_k, \mathbf{x}_{k-1} \mid \mathbf{z}_{1:k-1})$, whereas $\{\bar{\mathbf{y}}_k^{(i)}\}_{i=1}^N$, $\bar{\mathbf{y}}_k^{(i)} = [(\bar{\mathbf{x}}_k^{(i)})^T, (\mathbf{x}_{k-1}^{\prime(i)})^T]^T$ simulates $q(\bar{\mathbf{y}}_k) = \mathcal{N}(\bar{\mathbf{y}}_k \mid \boldsymbol{\mu}_k, \boldsymbol{\Sigma}_k)$, with mean $\boldsymbol{\mu}_k$ and covariance $\boldsymbol{\Sigma}_k$ where

$$\boldsymbol{\mu}_k = N^{-1} \sum_{i=1}^{N} \bar{\mathbf{y}}_k^{(i)}, \quad \boldsymbol{\Sigma}_k = N^{-1} \sum_{i=1}^{N} \left[\bar{\mathbf{y}}_k^{(i)} - \boldsymbol{\mu}_k\right] \left[\bar{\mathbf{y}}_k^{(i)} - \boldsymbol{\mu}_k\right]^T$$

4: **for** $i = 1, \dots, N + N_{Burn-in}$ **do**
5: Draw $(\bar{\mathbf{x}}_k, \mathbf{x}_{k-1}) \sim q(\bar{\mathbf{y}}_k)$.
6: Accept the new move as a sample in the chain $\mathbf{x}_k^{(i)} = \bar{\mathbf{x}}_k$ with probability α given in (8.11).
7: **end for**
8: Retain only N samples $\mathbf{x}_k^{(i)}$ subsequent to the end of the burn-in period.

Algorithm 8.2. Alternate steering MCMC.

1: Simulate $q_1(\bar{\mathbf{y}}_k)$ and $q_2(\bar{\mathbf{y}}_k)$ using two distinct steering distributions p_λ^1 and p_λ^2, respectively.
2: **for** $i = 1, \dots, N + N_{Burn-in}$ **do**
3: Draw $(\bar{\mathbf{x}}_k, \mathbf{x}_{k-1}) \sim q(\bar{\mathbf{y}}_k)$ where $q(\bar{\mathbf{y}}_k) = q_1(\bar{\mathbf{y}}_k)$ if $(i \bmod 2) = 1$, and $q(\bar{\mathbf{y}}_k) = q_2(\bar{\mathbf{y}}_k)$, otherwise.
4: Accept the new move as a sample in the chain $\mathbf{x}_k^{(i)} = \bar{\mathbf{x}}_k$ with probability α given in (8.11).
3: **end for**
4: Retain only N samples $\mathbf{x}_k^{(i)}$ subsequent to the end of the burn-in period.

of conventional particle filters such as the SIR-PF, as those consistently failed to converge in the underlying settings). The MCMC schemes use no more than 4000 particles and 2000 burn-in samples. The steering distributions p_λ^1 and p_λ^2 are set as $U[-6, 6]$ and $U[0, 6]$, respectively. The nonalternating MCMC scheme uses $\lambda \sim p_\lambda^1$.

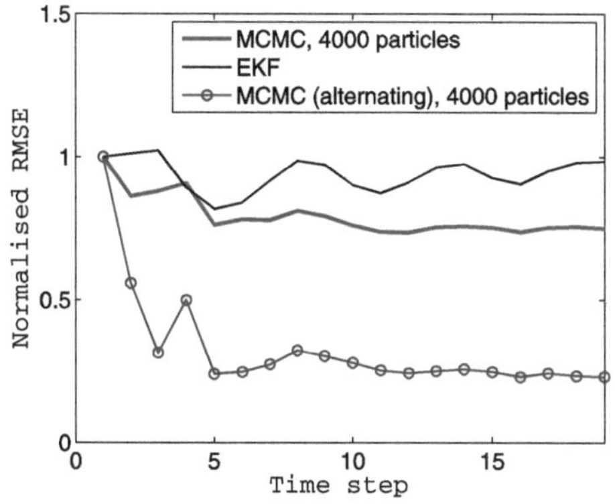

(a) Filtering performance

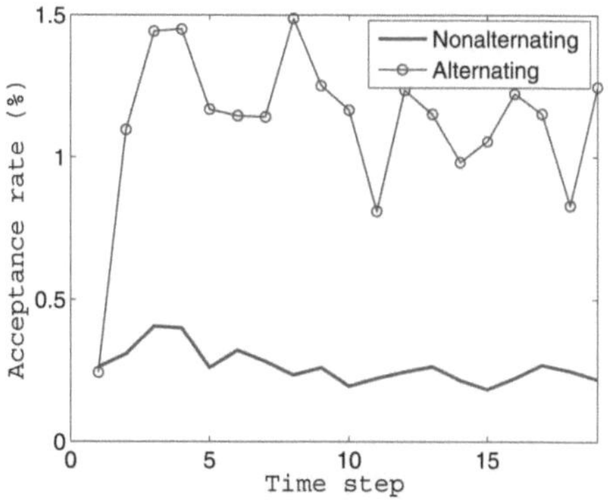

(b) Mean acceptance rate

Figure 8.1. RMSE of the various filtering schemes (left panel) and the corresponding mean acceptance rate of the MH (right panel), computed based on 20 Monte Carlo runs. Nonlinear system with 100 states.

In the first example we consider a system model which is an extension of the wide spread example from [1]:

$$\mathbf{x}_k^j = \mathbf{x}_{k-1}^j + \frac{25\sum_{i=1}^{n_x} \mathbf{x}_{k-1}^i}{1+\left(\sum_{i=1}^{n_x} \mathbf{x}_{k-1}^i\right)^2} + \cos(1.2k) + \mathbf{v}_{k-1}^j \tag{8.12a}$$

$$\mathbf{z}_k^j = \frac{\left(\mathbf{x}_k^j\right)^2}{20} + \mathbf{r}_k^j, \quad j = 1, \ldots, n_x, \tag{8.12b}$$

where the superscript j denotes the j-th element in the vector. The noises $\mathbf{v}_k^j$ and $\mathbf{r}_k^j$ are assumed to be Gaussian. All other related parameters are set as in [1].

The performance of the EKF and of both MCMC variants, namely, the alternate steering MCMC (alternating) and the nonalternating one, is shown in Figure 8.1. The normalised RMSE, defined as $E\,[\|\,\hat{\mathbf{x}}_k - \mathbf{x}_k\,\|_2 \,/\, \|\,\mathbf{x}_k\,\|_2]^{1/2}$, is approximated based on 20 Monte Carlo runs and is shown in Figure 8.1a. Figure 8.1b presents the mean acceptance rate of both MCMC variants. Figure 8.1 demonstrates the superiority of the alternating MCMC filtering approach in terms of estimation accuracy and sampling efficiency (i. e., improved acceptance rate).

In our last example we demonstrate the performance of the nonalternating MCMC approach in solving a dynamic compressed sensing problem [7]. The system model is similar to the one considered in [7] with $n_x = 100$ states, of which only 10 are either nonvanishing or noncompressible (i. e., the state process is sparse/compressible).

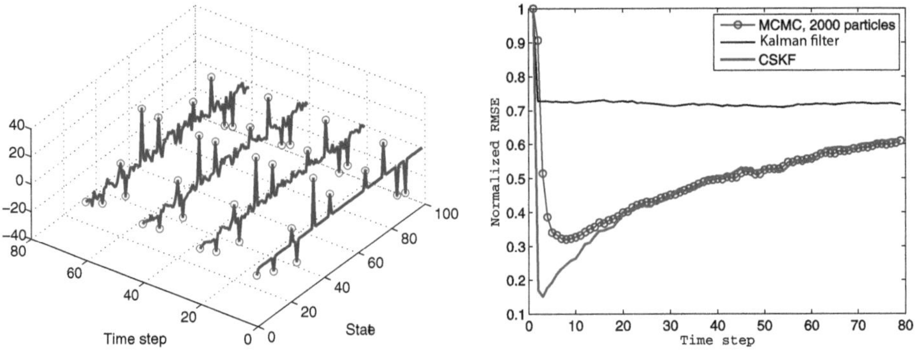

(a) Signal (b) Filtering performance

Figure 8.2. Dynamic compressed sensing example. A typical signal realization is shown over time (upper left panel). The normalized RMSE of the various filtering schemes, the Kalman filter, the CSKF, and the MCMC, is shown in the bottom panel. System with 100 states.

The signal itself becomes corrupted over time, and its corresponding complexity in the sense of sparseness rises (illustrated in Figure 8.2). At each time step there are 20 observations obeying $\mathbf{z}_k = H\mathbf{x}_k + \mathbf{r}_k$. The likelihood used by the MCMC algorithm is given as

$$p(\mathbf{z}_k \mid \mathbf{x}_k) \propto \exp\left\{-0.5 \parallel \mathbf{z}_k - H\mathbf{x}_k \parallel_2^2 - \beta \parallel \mathbf{x}_k \parallel_1\right\}, \tag{8.13}$$

with $\beta = 100$. The MCMC algorithm uses no more than 2000 particles and 2000 burn-in samples. The performance shown in Figure 8.2 demonstrates the viability of the MCMC approach in solving the compressed sensing problem. The estimation errors of both the CSKF and the MCMC nearly coincide from a certain compressibility level (the increased complexity is manifested by the positive slope of the estimation error lines). The estimation accuracy of a conventional Kalman filter is also depicted showing its uselessness for such a problem.

8.5 Conclusions

This chapter presents a Markov-chain Monte Carlo approach for high-dimensional nonlinear filtering. The new algorithm utilizes an improved proposal distribution which essentially incorporates the subgradient information of the underlying likelihood. This proposal is then used for generating candidate moves in high-probability regions of the sample space. A numerical study shows the potential of the new filtering scheme for high-dimensional nonlinear state estimation.

Acknowledgments. We acknowledge the support of the European Community's Seventh Framework Programme [FP7/2007-2013] under grant agreement No 238710. (Monte Carlo based Innovative Management and Processing for an Unrivalled Leap in Sensor Exploitation).

References

[1] M. S. Arulampalam, S. Maskell, N. Gordon, and T. Clapp, A Tutorial on Particle Filters for Online Nonlinear/Non-Gaussian Bayesian Tracking, *IEEE Trans. on Signal Proc.* **50** (2002), 174–188.

[2] A. Carmi, F. Septier, and S. J. Godsill, The Gaussian Mixture MCMC Particle Algorithm for Dynamic Cluster Tracking, *Automatica*, (2011).

[3] J. Liu and R. Chen, Sequential Monte Carlo Methods for Dynamic Systems, *Journal of the American Statist. Association* **93** (1998), 1032–1044.

[4] S. K. Pang, J. Li, and S. J. Godsill, Detection and Tracking of Coordinated Groups, *IEEE Transactions on Aerospace and Electronic Systems*, **47** (2011), 472–502.

[5] C. Andrieu, A. Doucet, and R. Holenstein, Particle Markov chain Monte Carlo methods, *Journal of the Royal Statistical Society: Series B (Statistical Methodology)* **74** (2010), 269–342.

[6] C. Berzuini, G. Nicola, W. R. Gilks, and C. Larizza, Dynamic Conditional Independence Models and Markov Chain Monte Carlo Methods, *J. Amer. Statist. Assoc.* **92** (1997), 1403–1412.

[7] A. Carmi, P. Gurfil, and D. Kanevsky, Methods for Sparse Signal Recovery Using Kalman Filtering With Embedded Pseudo-Measurement Norms and Quasi-Norms, *IEEE Transactions on Signal Processing* **58** (2010), 2405–2409.

[8] F. Daum and J. Huang, Curse of dimensionality and particle filters, in: *Proc. IEEE Aerospace Conf.*, vol. 4, pp. 1979–1993, 2003, DOI 1109/AERO.2003.1235126

[9] F. Daum and J. Huang, Particle flow for nonlinear filters with log-homotopy, in: *Proc. SPIE conference on Signal and Data Processing of Small Targets*, vol. 6969, article 696918, 2008, DOI: 10.1117/12.764909.

[10] N. Z. Shor, *Minimization Methods for Non-differentiable Functions*, Springer, Berlin Heidelberg New York Tokyo, 1985.

Author information

Avishy Carmi, Department of Mechanical and Aerospace Engineering, Nanyang Technological University, Singapore.
Email: `acarmi@ntu.edu.sg`

Lyudmila Mihaylova, School of Computing and Communications, Lancaster University, UK.
Email: `mila.mihaylova@lancaster.ac.uk`

Chapter 9

Game-Method for Modeling and WRF-Fire Model Working Together

Nina Dobrinkova, Stefka Fidanova, Ivan Dimov, Krassimir Atanassov, and Jan Mandel

Abstract. In this chapter we will evaluate the wind field distribution in case of wildland fire. In the WRF-fire model and the Game-Method for modeling we consider the wind as an average approximation with a certain direction. In reality this is not the case. The wind is nonstable and has unexpected changes in direction. That is why we decided to consider an approach which represents the wind as a random field with stochastic approximation. In this cbapter we will describe the possibility for implementation of Monte Carlo (MC) methods for wind-field variations in real wildland fire simulation.

Keywords. Monte Carlo, Game Model,WRF-Fire Model.

Mathematics Subject Classification 2010. 68U35, 68Q80, 97R50.

9.1 Introduction

Wildland fires are a problem for the southern member states of the European Union. In the recent years many statistics have been collected in Europe which seem to show that in the future, together with the climate change, this natural hazard will increasingly occur . Bulgaria is no exception. In the last 30 years forest fires in Bulgaria have increased, with a drastic increase 2000. In the late afternoon of July 23, 2003 [1] in the protected zone of the Pirin national park a fire occurred in a forest of a special type black-needle trees, whose wood is very. The fire burned 600 hectares, whereby three members of the military team participating in the flight of MI-17 helicopter died. The reason for these casualties was that the helicopter was flying too low above the burning fire, where the amount of oxygen is very low, with the result that the helicoptor motor could not work properly. Also the forest fires are affecting local meteorological conditions while burning, of which fact the helicopter team was not aware. These factors caused the loss of life. In 2007 according to civil protection reports [2], in some places extremely high temperatures of more than 45 °C have been measured during July, causing large fires in Sofia, Sliven, Smolian, Bourgas, and Stara Zagora. the one close to Stara Zagora estimated to be most extensive. More than 30,000 hectares were affected in this mountain region. The reason for this was the steep slopes of the burning area, where it is impossible to use standard firefighting equipment. That summer there were no casualties. These fires were caused by human errors during the use of fire in open spaces, in combination with weather conditions. The burned hectares are are a huge loss, but the human death toll is a loss which cannot be measured. This is why our team believes that research in wind behavior in the case of wildland fires can significantly reduce the negative statistical damage caused by these extreme events.

In this chapter we combine the Game-Method for modeling with WRF-Fire mode, and we include the wind in two manners: i) average value for the wind, and ii) random field describing the wind. The chapter is organized as follows. In Section 9.2 the Game-Method for modeling is described. In Section 9.3 the WRF-Fire model and its meteorological aspect is explained. In Section 9.4 we introduce the combination of the two methods and two manners of wind including. In Section 9.5 some conclusions and directions for future work are given.

9.2 Description of the Game-Method for Modeling

In a series of papers the idea for the Game-Method for modeling has been described (see [14–16]). It is, briefly, as follows. Let us have n- (in our case: 2-)dimensional space covered by n-dimensional cubes (in our case: squares). Conway's game of life contains the same plane with the same cover, but in his case, the cells (squares) contain or do not contain asterisks. Here the asterisks are replaced with symbols from some fixed set S. On the one hand, the symbols correspond to the existence or absence of asterisks, and on the other hand, they show an additional quality of the objects (if they exist), which they represent. In addition, there are now axioms (rules) for movement of the symbols from a cell to one of its neighbors, and axioms (rules) for inference of two or more symbols which a a result of movement enter a cell.

For example, if we want to model wildland dynamics, we will put into cells digits representing the burning coefficient of the area. The rules for movement of the symbols will represent the changing situation in the cell, the rate of burning.

Let A be the set of axioms. Let us have some configuration K representing places of symbols in a fixed finite (or infinite) place. Let the result of applying the rules from A on the symbols from K be noted by $A(K)$. Let L be the final configuration obtained by A from K, i. e., $L = A(K)$. The game method for modeling proposes that we can check the validity of the axioms A on the basis of a sufficiently large number of initial configurations $K_1, K_2, \ldots, K_n$ that generate a respective set of final configurations $L_1, L_2, \ldots, L_n$. On the basis of the initial and final configurations, we determine the average initial (K^*) and average final (L^*) configurations. Some criteria for proximity of two configurations are discussed in [14, 16]. When configurations $A(K^*)$ and L^* are close according to some of these criteria, we can assert that the axioms A are constructed in a (statistically) correct way. Thus we can check the validity of the axioms, representing a given process. If the axioms of A are correct, we can extend their list with new ones and the check will be repeated.

9.3 General Description of the Coupled Atmosphere Fire Modeling and WRF-Fire

In this section we focus on atmosphere-fire modeling in WRF-Fire. It is based on [3], where more details can be found.

Fire models range from simple spread formulas to sophisticated computational fluid dynamics and combustion simulations (see [3, p. 50]). However, a fire behavior model in a decision support system should be faster than real time in order to deliver a prediction, which dictates a compromise between the spatial resolution, the processes to be modeled, and fast execution.

Weather has a major influence on wildfire behavior; in particular, the fire spread speed is largely determined by the wind. Conversely, the fire influences the weather through the heat and vapor fluxes. Fire heat output can easily reach the surface intensity of $1\mathrm{MW/m^2}$, and the fast-rising hot air causes significant air motion, which also affects the atmosphere away from the fire. It is known that a large fire "creates its own weather." The correct fire shape and progress result from the two-way interaction of the fire and the atmosphere [5, 6].

WRF-Fire [3] combines the Weather Research and Forecasting Model (WRF) [7] with a semiempirical fire spread model from [5], based on a modified Rothermel's formula, which approximates the fire spread rate in the normal direction to the fireline as a function of fuel properties, wind speed close to the ground, and terrain slope. The semiempirical model approximates the rate of the fuel decay through burning by an exponential in time. The semiempirical formulas were derived from laboratory experiments, and the coupled model was verified on several large fires [6] in an earlier implementation, with the fire propagation by tracers and atmospheric modeling by the Clark–Hall weather code. WRF-Fire takes advantage of this validation and implements a subset of the physical model from [5, 6]. The fire spread in WRF-Fire is implemented by the level-set method [8]. The level-set function can be manipulated more easily than tracers. The weather model has been replaced by WRF, a supported standard community weather code. WRF can be run with several nested refined meshes, called domains in meteorology, which can run different physical models. The fundamental observation here is that the innermost domain, which interacts directly with the fire model, needs to run in the Large Eddy Simulation (LES) mode [9]. WRF-Fire takes advantage of the mature WRF infrastructure for parallel computing and for data management. An important motivation for the development of the WRF-Fire software was the ability of WRF to export and import states, thus facilitating data assimilation (input of additional data while the model is running), which is essential for fire behavior prediction from all available data [10].

WRF contains the WRF Preprocessing System (WPS) [11, Chap. 3], which can input meteorology and land use data in a large number of commonly used formats. WPS has been extended to process fine-scale land data for use with the fire model, such as topography and fuel information in [12] and [11, App. A]. While the format of meteorology data has largely stabilized, the ingestion of fire-modeling data was developed for US sources only, and it will require further modifications or preprocessing for other countries.

Mathematically, the fire model is posed in the horizontal (x, y) plane. The semi-empirical approach to fire propagation used here assumes that the fire spreads in the direction normal to the fireline at the spin given by the modified Rothermel's formula

$$S = \min\{B_0, R_0 + \phi_W + \phi_S\},$$

where B_0 is the backing rate (spread rate against the wind), R_0 is the spread rate in the absence of wind, $\phi_W = a(\vec{v} \cdot \vec{n})^b$ is the wind correction, and $\phi_S = d\nabla z \cdot \vec{n}$ is the terrain correction. Here $\vec{v}$ is the wind vector, ∇z is the terrain gradient vector, and $\vec{n}$ is the normal vector to the fireline in the direction away from the burning area. In addition, the spread rate is limited by $S \leq S_{\max}$. Once the fuel is ignited, the amount of the fuel at location (x, y) is given by

$$F(x, y, t) = F_0(x, y)e^{-(t-t_i(x,y))/T(x,y)} \qquad t > t - t_i(x, y), \tag{9.1}$$

where t is the time, t_i is the ignition time, F_0 is the initial amount of fuel, and T is the time constant of fuel (the time for the fuel to burn down to $1/e$ of the original quantity). The coefficients $B_0, R_0, a, b, d, S_{\max}, F_0$, and T are data.

The heat flux from the fire is inserted into the atmospheric model as forcing terms in differential equations of the atmospheric model in a layer above the surface, with exponential decay with altitude, using the code from [9]. The sensible heat flux is inserted as the time derivative of the temperature, while the latent heat flux as the derivative of water vapor concentration. This scheme is required because atmospheric model with explicit timestepping, such as WRF, do not support flux boundary conditions. The heat fluxes from the fire to the atmosphere are taken proportional to the fuel burning rate, $\partial F(x, y, t)/\partial t$. The proportionality constants are again fuel coefficients.

For each point in the plane, the fuel coefficients are given by one of the 13 Anderson categories [13]. The categories are developed for the US and assume wind measured at a certain altitude. WRF-Fire provides for the definition of the categories as input data and the specification of the altitude of the wind input into the spread formula, which allows the software to adapt to other countries.

The burning region at time t is represented as the set of all points (x, y) where a level set function $\phi(x, y, t) < 0$. The level set function satisfied the partial differential equation,

$$\partial\phi/\partial t = -S\,|\nabla\phi|, \tag{9.2}$$

where $|\nabla\phi|$ is the Euclidean norm of the gradient [8].

In each time of the atmospheric model, the winds are first interpolated from the atmospheric model grid to a finer fire model grid. A numerical scheme for the level set equation (9.2) is then advanced to the next time step value, and the fuel burned during the time step is computed by quadrature from (9.1) in each fire model cell. The resulting heat fluxes are averaged over the fire cells that make up one atmosphere model cell.

9.4 Wind Simulation Approach

The main goal of this work is the combination of the Game-Method for modeling and the WRF-Fire model. Thus we will include the meteorology in the model, which is very important. In WRF-Fire the wind is included by its average wind value. Other way is to represent the wind by a random field and to use Monte Carlo simulation.

Monte Carlo simulation is becoming a fundamental tool for the design of complex and important wind-excited structures. A common application regards the time-domain dynamic analysis of multidot nonlinear structures whose excitation is calculated on the base of simulated wind velocity time histories. For the two models described we suppose that we have two ways of including the wind parameter in the models. The first one is when we have the average value for the wind in conditions of fixed direction and velocity. This is a simplified but very commonly used way to include the wind parameter, and it i used in the meteorological calculations in models such as MM5. This case is not very close to the reality because by its nature the wind has an impulse character. It can not perform as a constant front blowing only in one or other direction.

The second approach is the usage of a random field describing the wind, which is a much more realistic representation of the natural behavior of the wind. A random field is a collection of random variables defined by their joint distribution functions. Now we define $Z(x)$, $x \in R^2$ to be a (spatial) random field, and then we have:

$$F_{x_1,x_2,\ldots,x_n}(X_1,\ldots,X_n) = P(z_1 \leq X_1,\ldots,z_n \leq X_n),$$

where $z_1 = Z(x_1)$ and n can be any integer.

For this second case we denote that it can be split into two more subcases. The first one is when the direction and the velocity of the wind have a random behavior (turbulence). This case is very complicated and difficult for computational estimations. For the second subcase we denote that the wind has a fixed direction and random velocity. The case is simplified, and we will focus on this in our research work. When we treat the turbulence the wind field is over any nonsmooth area and can be decomposed into two components: steady state flow (i. e. the average flow over a sufficiently long time interval) and turbulent flow (i. e. the difference between the steady state flow and the observed flow). Typically the steady state flow can be modeled using the equations of motion, the simplest form of which is the geostrophic approximation – i. e., by models based on the integration of the differential equations governing fluid motion. In practice these are regarded as random and are modeled statistically.

Let us set that the width of the random field as parameter w, which will have a central role in our algorithm. We will simulate the behavior of the random field width w when it has different values. In cases where the values are small, this can be a base that the wind speed variation will be in a small interval similar to the case where the velocity is fixed. In the case where w has large value relativity the random field of the wind can be different in range values. The reason for this could be the nonstable and nonconstant nature of wind behavior, and we can calibrate the models in correlation with the meteorological conditions. In the described steps with conditions for the random field of the wind, Monte Carlo methods can be used for simulation results and tests. Our intention here is to compare the exact rotational-wind estimator, its two-dimensional version, and the approximate estimator. Wind velocity is usually idealized as the sum of two contributing parts: a mean part, assumed as invariant within a time interval of conventional length, and a fluctuating part, identified with the atmospheric turbulence and modeled as a stationary Gaussian random process. The intensity and direction of the mean wind velocity varies with the position in the atmospheric boundary layer, depending on the ground roughness, the topography properties, the atmospheric stratification and the geographic latitude. The statistical properties of turbulence depend on the above parameters and on the intensity of the mean wind velocity. Such parameters identify a wind state, for which the probability of occurrence is determined by means of on-site measurements.

9.5 Conclusion

The present chapter describes a scheme for the representation of the two- and three-dimensional turbulence over discretized spatial domains. We propose a combination of the game modeling method and meteorology from the WRF-Fire method. The wind is simulated by the random field and Monte Carlo methods. Our future work will be focused on learning the random field parameters in various range for w. We will test by sensitivity analysis the influence of the value of this parameter on the models' output. We will compare the achieved results in the cases of fixed velocity.

Acknowledgments. The authors would like to acknowledge the support provided by the Bulgarian National Science Fund under Grants Ref. No. DID-02-29 "Modeling Processes with Fixed Development Rules" and DMU-02-14.

References

[1] http://news.ibox.bg/news/id_1888536796 (last accessed September 21, 2012).

[2] http://ec.europa.eu/echo/civil_protection/civil/forestfires_bg_2007.htm (last accessed September 21, 2012).

[3] J. Mandel, J. D. Beezley, J. L. Coen, and M. Kim, Data assimilation for wildland fires: Ensemble Kalman filters in coupled atmosphere-surface models. *IEEE Control Systems Magazine* **29** (2009) 47–65.

[4] A. L. Sullivan, A review of wildland fire spread modelling, 1990–present, 1: Physical and quasi-physical models, 2: Empirical and quasi-empirical models, 3: Mathematical analogues and simulation models, *International Journal of WildLand Fire* **18** (2009) 1: 347–368, 2: 369–386, 3: 387–403.

[5] T. L. Clark, J. Coen, and D. Latham, Description of a coupled atmosphere-fire model, *International Journal of Wildland Fire* **13** (2004) 49–64.

[6] J. L. Coen, Simulation of the Big Elk Fire using coupled atmosphere-fire modeling, *International Journal of Wildland Fire* **14**(1) (2005) 49–59.

[7] WRF Working Group, Weather Research Forecasting (WRF) Model, http://wrf-model.org (last accessed July 2010).

[8] S. Osher and R. Fedkiw, *Level set methods and dynamic implicit surfaces,* Springer, New York, 2003.

[9] E. G. Patton and J. L. Coen, WRF-Fire: A coupled atmosphere-fire module for WRF, in: Preprints of Joint MM5/Weather Research and Forecasting Model Users' Workshop, Boulder, CO, June 22–25. NCAR, pp. 221–223, 2004, available online at http://www.mmm.ucar.edu/mm5/workshop/ws04/Session9/Patton_Edward.pdf (last accessed September 21, 2012).

[10] J. Mandel, M. Chen, L. P. Franca, C. Johns, A. Puhalskii, J. L. Coen, C.C. Douglas, R. Kremens, A. Vodacek, and W. Zhao, A note on dynamic data driven wildfire modeling, in: M. Bubak, G. D. van Albada, P. M. A. Sloot, and J. J. Dongarra (eds.), *Computational Science – ICCS 2004*, pp. 725–731, Lecture Notes in Computer Science 3038, Springer, Heidelberg Berlin, 2004.

[11] W. Wang, C. Bruyère, M. Duda, J. Dudhia, D. Gill, H. C. Lin, J. Michalakes, S. Rizvi, X. Zhang, J. D. Beezley, J. L. Coen, and J. Mandel, *ARW version 3 modeling system user's guide*, Mesoscale and Miscroscale Meteorology Division, National Center for Atmospheric Research, 2010, available online at http://www.mmm.ucar.edu/wrf/users/docs/user_guide_V3/ARWUsersGuideV3.pdf (last accessed September 21, 2012).

[12] J.D. Beezley, How to run WRF-Fire with real data, http://www.openwfm.org/wiki/How_to_run_WRF-Fire_with_real_data (last accesed July 2010).

[13] H.E. Anderson, Aids to determining fuel models for estimating fire behavior, USDA Forest Service, Intermountain Forest and Range Experiment Station, Research Report INT-122, 1982, http://www.fs.fed.us/rm/pubs_int/int_gtr122.html (last accessed September 21, 2012).

[14] K. Atanassov, On a combinatorial game-method for modelling, *Advances in Modelling & Analysis, AMSE Press* **19**(2) (1994), 41–47.

[15] L. Atanassova and K. Atanassov, *Intuitionistic fuzzy interpretations of Conway's game of life*, pp. 232–239, Lecture Notes in Computer Science 6046, Springer, Berlin, 2011.

[16] L. Atanassova and K. Atanassov, On a game-method for modelling with intuitionistic fuzzy estimations, Part 2, this volume..

Author information

Nina Dobrinkova, Institute of Information and Communication Technologies, Bulgarian Academy of Sciences, Sofia, Bulgaria.
Email: nido@math.bas.bg

Stefka Fidanova, Institute of Information and Communication Technologies, Bulgarian Academy of Sciences, Sofia, Bulgaria.
Email: stefka@parallel.bas.bg

Ivan Dimov, Institute of Information and Communication Technologies, Bulgarian Academy of Sciences, Sofia, Bulgaria.
Email: ivdimov@bas.bg

Krassimir Atanassov, Institute of Biophysics and Biomedical Engineering, Bulgarian Academy of Sciences, Sofia, Bulgaria.
Email: krat@bas.bg

Jan Mandel, Department of Mathematical and Statistical Sciences, University of Colorado, Denver, CO, USA.
Email: jan.mandel@ucdenver.edu

Chapter 10

Wireless Sensor Network Layout

Stefka Fidanova, Pencho Marinov, and Enrique Alba

Abstract. Wireless Sensor Networks (WSN) allow the monitoring of wide and remote areas with precision and liveness unseen to date without the intervention of a human operator. The evolution of wireless networking technologies and their key role in future internet scenarios offer an increasing wealth of opportunities for distributing data over wireless networks. When deploying a WSN, the major objective is to achieve full coverage of the terrain. Other objectives are the minimum number of sensor nodes and energy efficiency of the network. In this chapter we address a WSN layout problem instance in which full coverage is treated as a constraint, while objective function is a product of the number of the sensors and energy efficiency. To solve this we propose the Ant Colony Optimization (ACO) algorithm. The terrain is modeled with a 500×500-point grid, and both sensing radius and communication radius are set to 30. We compare our results with existing evolutionary algorithms.

Keywords. Wireles Sensor Network, Monte Carlo Method, Ant Colony Optimization.

Mathematics Subject Classification 2010. 90b18, 90B50, 68M10, 90C29.

10.1 Introduction

Telecommunications is a general term for a vast array of technologies that send information over distances. Mobile phones, land lines, satellite phones- and voice-over internet protocol are all telephone technologies – altogether one single field of telecommunications. Radio, television, and networks are a few more examples of telecommunication. Nowadays, the trend in telecommunication networks is to have highly decentralized, multinode networks. From small, geographically close, size-limited local area networks, the evolution has led to the huge global internet. This same path has been followed in wireless communications, where we can already see wireless telephoning able to reach virtually any city in the world. Within this context WSNs have recently become a hot topic in research. A WSN allows an administrator to automatically and remotely monitor almost any phenomenon with a precision unseen until now. The use of multiple small cooperative devices yields a brand new horizon of possibilities, yet produces a great number of new problems which need to be solved. WSNs have so far been employed in military activities such es reconnaissance, surveillance, and target acquisition [4], in environmental activities such as forest fire prevention, in geophysical activities such as volcano eruptions study [12], for biomedical purposes such as health data monitoring [13], or in civil engineering [10].

The wireless sensors fulfill two fundamental functions: sensing and communicating. The sensing can be of different types (seismic, acoustic, chemical, optical, etc.), and the communication is performed wirelessly. However, the small size and energy storage capacity of the sensors prevent them from relaying their gathered information directly to the base. It is therefore necessary for them to transmit their data to a high-energy communication node (HECN) which is able to provide the transmission relay to an aircraft or a satellite. All sensors must be able to transmit their data to this node, either directly or via hops, using nearby sensors as communication relays.

When deploying a WSN, the positioning of the sensor nodes becomes one of the major concerns. The coverage obtained with the network and the economic cost of the network depend directly of it. Since many WSNs can have a large number of nodes, the task of selecting the geographical positions of the nodes for an optimally designed network can be very complex. Therefore, metaheuristics seems to be an interesting option for solving this problem.

In this chapter we propose a solution method for the WSN layout problem using ACO. We focus on both minimizing the energy depletion of the nodes in the network and minimizing the number of nodes, while the full coverage of the network and connectivity are considered as constraints.

Jourdan [5] solved an instance of WSN layout using a multiobjective genetic algorithm. In there formulation a fixed number of sensors had to be placed in order to maximize the coverage. In [8] several evolutionary algorithms were proposed for solving the problem. In [7] genetic algorithm was proposed which achieves similar solutions as the algorithms in [8], but it was tested on small test problems.

For solving the WSN layout problem, the coverage has to satisfy some restrictions, and the largest possible coverage is preferred: the number of sensor nodes should be kept low for economical reasons, and the network needs to be connected. Finally, the energy management of the network is a key issue which has to be taken into account, since the life span of the network depends on it.

The rest of this chapter is organized as follows. In Section 10.2 WSN is described and the layout problem is formulated. Section 10.3 presents the ACO algorithm. In Section 10.4 the experimental results obtained are shown. Finally, several conclusions are drown in Section 10.5.

10.2 Wireless Sensor Network Layout Problem

A wireless sensor network (WSN) consists of spatially distributed autonomous sensors to cooperatively monitor physical or environmental conditions, such as temperature, sound, vibration, pressure, motion, or pollutants [1, 6]. The development of wireless sensor networks was motivated by military applications such as battlefield surveillance and are now used in many areas of industrial and civilian applications, including industrial process monitoring and control, machine health monitoring, environment and habitat monitoring, healthcare applications, home automation, and traffic control [6].

In addition to one or more sensors, each node in a sensor network is typically equipped with a radio transceiver or other wireless communications device, a small microcontroller, and an energy source, usually a battery. A sensor node might vary in size from that of a box to the size of a grain of dust [6], although functioning "motes" of genuine microscopic dimensions have yet to be created. The cost of sensor nodes is similarly variable, ranging from a few cents to hundreds of dollars, depending on the size of the sensor network and the complexity required of individual sensor nodes [6]. Size and cost constraints on sensor nodes result in corresponding constraints on resources such as energy, memory, computational speed and bandwidth [6].

A sensor network normally constitutes a wireless ad hoc network, meaning that each sensor supports a multihop routing algorithm where nodes function as forwarders, relaying data packets to a base station.

A wireless sensor network is a wireless network formed by sensor nodes. Each sensor node senses an area around itself called its sensing area. A parameter called sensing radius determines the sensitivity range of the sensor node and thus the sensing area. The nodes communicate among themselves using wireless communication links. These links are determined by a communication radius. A special node in the WSN called high energy communication node (HECN) is responsible for external access to the network. Therefore, every sensor node in the network must have communication with the HECN. Since the communication radius is often much smaller than the network size, direct links are not possible for peripheral nodes. A multihop communication path is then established for those nodes that do not have the HECN within their communication range.

The WSN layout problem amounts to deciding the geographical position of the sensor nodes that form a WSN. In our formulation, a nonfixed amount of sensor nodes has to be placed in a terrain providing full sensitivity coverage. The positions of the nodes have to be chosen in a way that minimizes the energy spent in communications by any single node, while keeping the connectivity of the network. These are opposed objectives since the more nodes there are the lesser share of retransmissions they bear.

In order to determine the energy spent by communications, the number of transmissions which each node performs is calculated. The WSN operates by rounds: in a round, each node collects the data from its measurements and sends it to the HECN. Each node transmits the information packets to the neighbor closest to the HECN, or to the HECN itself if it is within the communication range. When several neighbors are tied for the shortest distance from the HECN, the traffic is distributed evenly among them using random assignment. That is, if a node has n neighbors tied for shortest distance from HECN, each one receives $1/n$ of its traffic load. Therefore, every node has a traffic load equal to 1 (corresponding to its own sent data) plus the sum of all traffic loads received from neighbors that are farther from the HECN. The sensing area of the WSN is the union of the individual areas of all nodes. The designer wants the network to cover the complete sensing area. On the other hand, the number of sensor nodes must be kept as low as possible, since using many nodes represents a high cost of the network, possibly influences the environment, and also provokes the probability of detection (when stealth monitoring is designed). The objective of this problem is to minimize network energy and the number of sensors deployed while the area is fully covered and connected.

10.3 ACO for WSN Layout Problem

Many of the existing solutions to this problem come from the field of evolutionary computation [2, 8]. After analyzing them, we noticed that these interesting developments are quite similar to ACO algorithms.

Real ants foraging for food lay down quantities of pheromone (chemical cues) marking the path they follow. An isolated ant moves essentially guided by a heuristic function, and an ant encountering a previously laid pheromone will with high probability detect and decide to follow it, thus taking more informed actions based on the experience of previous ants (and thereby reinforce it with a further quantity of pheromone). The repetition of the above mechanism represents the autocatalytic behavior of a real ant colony where the more ants follow a trail, the more attractive that trail becomes.

The ACO algorithm uses a colony of artificial ants that behave as cooperative agents in a mathematic space were they are allowed to search and reinforce pathways (solutions) in order to find the optimal ones. The problem is represented in a graph and the ants walk on the graph to construct solutions. The solution is represented by a path in the graph. After initialization of the pheromone trails, ants construct feasible solutions, starting from random nodes, and then the pheromone trails are updated. At each step ants compute a set of feasible moves and select the best one (according to some probabilistic rules based on a heuristic guided function) to carry out the rest of the tour. The transition probability p_{ij}, to chose the node j when the current node is i, is based on the heuristic information η_{ij} and on the pheromone trail level τ_{ij} of the move, where $i, j = 1, \ldots., n$:

$$p_{ij} = \frac{\tau_{ij}^{\alpha} \eta_{ij}^{\beta}}{\sum_{k \in allowed} \tau_{ik}^{\alpha} \eta_{ik}^{\beta}}. \tag{10.1}$$

The higher value of the pheromone and the heuristic information, the more profitable it is to select this move. In the beginning, the initial pheromone level is set to a small positive constant value τ_0, and then ants update this value after completing the construction stage [3]. ACO algorithms adopt different criteria to update the pheromone level.

In our implementation we use the MAX-MIN Ant System (MMAS) [11], which is one of the more popular ant approaches. The main feature of MMAS is to use a fixed upper bound $\tau_{\max}$ and a lower bound $\tau_{\min}$ of the pheromone trails. Thus the accumulation of big amounts of pheromone by part of the possible movements and repetition of same solutions is partially prevented.

The pheromone trail update rule is given by

$$\tau_{ij} \leftarrow \rho\tau_{ij} + \Delta\tau_{ij}, \tag{10.2}$$

$$\Delta\tau_{ij} = \begin{cases} 1/C(V_{best}) & \text{if } (i,j) \in \text{best solution} \\ 0 & \text{otherwise,} \end{cases}$$

where V_{best} is the iteration best solution and $i, j = 1, \dots, n$, $\rho \in [0, 1]$ models the evaporation in the nature, and $C(V)$ is the objective function. The WSN layout is problem a multiobjective problem, which we convert to a monoobjective one. The new objective function is a product of the number of sensors and the energy of the network, and we search for a solution which minimizes it. Thus when the energy and/or number of sensors decreases, the value of the objective function will decrease and the two components have equal influence. To avoid stagnation of the search, the range of possible pheromone values on each movement is limited to an interval $[\tau_{\min}, \tau_{\max}]$. $\tau_{\max}$ is an asymptotic maximum of τ_{ij}, and $\tau_{\max} = 1/(1-\rho)C(V^*)$, while $\tau_{\min} = 0.087\tau_{\max}$. Where V^* is the optimal solution, but it is unknown, we use V_{best} instead of V^*.

The WSN layout problem is represented by a graph as follows. The terrain is modeled by grid $G = \{g_{ij}\}_{N\times M}$; the pheromone is related with location sites $Ph = \{ph_{ij}\}_{N\times M}$, the initial pheromone can be a small value, for example $1/n_{\text{ants}}$. The central point, where the HECN is located, is included in the solutions as first point (zero point). Each ant begins to create the rest of the solution from a random node which communicates with the central one, and thus the different start of each ant in every iteration is guaranteed. The ant chooses the next position by the ACO probabilistic rule (Eq. 10.1). It choses the point having the higher probability. If there is more than one point with the same probability, the ant randomly chooses one of them. The construction of heuristic information is one of the crucial points of the ACO algorithm. The heuristic information needs to be constructed thusly so that the ants are better able to look for better solutions. For certain kinds of problems it is not obvious how to prepare this. One needs to combine different elements of the problem in a most appropriate way.

Our heuristic information is as follows:

$$\eta_{ij}(t) = s_{ij} l_{ij} (1 - b_{ij}), \tag{10.3}$$

where s_{ij} is the number of points which the new sensor will cover, and

$$l_{ij} = \begin{cases} 1 & \text{if communication exists} \\ 0 & \text{if there is no communication,} \end{cases} \tag{10.4}$$

where b is the solution matrix and the matrix element $b_{ij} = 1$ if there is a sensor on this position, otherwise $b_{ij} = 0$. With s_{ij} we try to locally increase the covered points; more new covered points eventually leads to a lower number of sensors. With l_{ij} we guarantee that all sensors will be connected; with rule $(1 - b_{ij})$ we guarantee that the position is not yet chosen and that no more than one sensor will be mapped onto the same position. When $p_{ij} = 0$ for all values of i and j, the search stops. Thus, the construction of the solution stops if there are no more free positions, if all points are covered, or if new communication is impossible.

10.4 Experimental Results

With this work we wish to contribute to the improvement the state-of-the-art use of metaheuristics for solving the WSN layout problem. Our aim is to provide an efficient solving method by comparing a set of state-of-the-art metaheuristic techniques applied in the same scenario. We want to solve a new flexible instance in which for the first time (to the best of our knowledge) both the number and positions of the sensors can be freely chosen, with full coverage of the sensor field guaranteed, while treating the energy efficiency and the overall cost of the network. Beyond this, our interest is to tackle complex instances in which the WSN size is in the same order of magnitude as a real WSN, with several hundred nodes.

With our algorithm we can solve WSN layout problems on any rectangular area. In this work we solve an WSN problem instance where a terrain of 500×500 m needs to be covered using nodes with coverage and communication radii of 30 m. The terrain has an area of 250,000 m^2, and each sensor covers 2827 square meters, meaning that in ideal conditions only 89 would be necessary. In fact, these ideal conditions do not exist, since this would imply that no overlap exists between any two node sensing areas, which is impossible due to their geometrical shape (circle). Therefore, the expected minimum number of nodes for full-coverage is higher than 89. An example of a solution that achieves full coverage of the region is a square grid formed by the sensors separated by 30 m. Starting at the HECN, 250 m have to be covered on each side of the terrain, requiring 8 sensors. Therefore the grid has 17 $(8 + 8 + 1)$ rows and 17 columns, thus 289 sensors including the HECN. In this symmetrical configuration there are 4 nodes directly connected to the HECN, so that in the complete traffic of the network 288 messages per round is evenly divided among them. This results in the most loaded nodes having a load of 72 messages. So this solution candidate obtains (288, 72). We call it symmetric solution.

We apply MAX-MIN ant algorithm with the following parameters: $\alpha = \beta = 1$, $\rho = 0.5$, the number of used ants is 3 and the maximum number of iterations is 10. In Table 10.1 the best found results (with respect to the sensors and with respect to the energy) are reported, achieved using several metaheuristic methods. We compare our ACO algorithm results with results obtained by the evolutionary algorithms in [8],

Table 10.1. Experimental results.

Algorithm	*min sensors*	*min energy*
symmetric	(288,72)	(288,72)
MOEA	(260,123)	(291,36)
NSGA-II	(262,83)	(277,41)
IBEA_{HD}	(265,83)	(275,41)
ACO	(233,58.8)	(239,58.2)

called MOEA, NSGA-II, IBEA, and the symmetric solution. These algorithms performs multiobjectively and reports nondominated solutions. One solution dominates another if all the components of the first are better than those of the second. A solution is nondominated if and only if no other solution dominates it.

We perform 30 independent runs of the ACO algorithm, and the achieved numbers of sensors are in the interval $[232, 247]$. Thus the worst number of sensors achieved by ACO algorithm is less than the best number of sensors achieved by the other algorithms mentioned. We observe that the achieved by MOEA solutions do not dominate the symmetric solution. Some of the achieved by NSGA-II and IBEA_{HD} solutions dominate the symmetric solution, but they do not dominate the MOEA solutions. The ACO algorithm achieves solutions with a lower number of sensors than other mentioned algorithms. The ACO solutions dominate part of the nondominated solutions of the other methods (the solutions with minimal number of sensors), but it does not dominate the solutions with minimal energy. In this case ACO achieves solutions with a lower number of sensors but with a greater value for energy. We can conclude that our ACO algorithm achieves very encouraging solutions.

10.5 Conclusion

We have defined the wireless sensor networks layout problem with its connectivity constraint. A very large instance consisting of a $500 \times 500\ \text{m}^2$ area has to be covered, using sensors nodes whose sensing and communication radii are 30 m in a way that minimizes both the number of sensors and the traffic load in the most loaded sensor node. We convert the problem from a multiobjective to a single objective one. The new objective function is a product of the number of sensors and energy of the network. Thus both "previous" objective functions have equal influence in the "new" objective function. We propose ACO algorithm for solving this problem, and we compare it with existing evolutionary algorithms. The results of the experiments has shown a very encouraging performance of the ACO algorithm.

Acknowledgments. This work has been partially supported by the Bulgarian National Scientific Fund under the grants DID 02/29 and DTK 02/44, and by Spanish Ministry of Science and Innovation and FEDER under contract TIN2008-06491-C04-01. It has also been partially funded by the Andalusian Government under contract P07-TIC-03044.

References

[1] I. F. Akuildiz, W. Su, Y. Sankarasubramaniam, and E. Cayrci, Wireless Sensor Networks: A Survey, *Computer Networks* **38**(4), (2001), 393–422.

[2] E. Alba and G. Molina, Optimal Wireless Sensor Layout with Metaheuristics: Solving a Large Scale Instance, in: *Large-Scale Scientific Computing*, pp. 527–535, LNCS 4814, Springer, Berlin Heidelberg, 2008.

[3] E. Bonabeau, M. Dorigo, and G. Theraulaz, *Swarm Intelligence: From Natural to Artificial Systems*, Oxford University Press, 1999.

[4] K. Deb, A. Pratap, S. Agrawal, and T. Meyarivan, A Fast and Elitist Multiobjective Genetic Algorithm: NSGA-II, *IEEE Transactions on Evolutionary Computation* **6**(2) (2002), 181–197.

[5] D. B. Jourdan, Wireless Sensor Network Planing with Application to UWB Localization in GPS-denied Environments, PhD thesis, Massachusetts Institut of Technology, 2000.

[6] R. Kay and M. Friedmann, The Design Space of Wireless Sensor Networks, *IEEE Wireless Communications* **11**(6) (2004), 54–61.

[7] A. Konstantinidis, K. Yang, Q. Zhang, and D. Zainalipour-Yazti, A multi-objective Ecolutionary Algorithm for the deployment and Power Assignment Problem in Wireless sensor Networks, *J. of Computer Networks* **54**(6) (2010), 960–976.

[8] G. Molina, E. Alba, and El-G. Talbi, Optimal Sensor Network Layout Using Multi-Objective Metaheuristics, *Universal Computer Science* **14**(15) (2008), 2549–2565.

[9] J. Nemeroff, L. Garcia, D. Hampel , and S. DiPierro, Application of Sensor Network Communications, in: *IEEE Military Comm. Conf.* pp. 336–341 2001.

[10] J. Paek, K. Chintalapudi, R. Govindan, J. Caffrey, and S. Masri, A wireless sensor network for structural health monitoring: performance and experience, in: *Proceedings of the 2nd IEEE workshop on Embedded Networked Sensors (EmNets'05)*, pp. 1–9. IEEE Computer Society, Washington, DC, 2005.

[11] T. Stutzle and H. H. Hoos, MAX-MIN Ant System, *Future Generation Computer Systems* **16** (2000), 889–914.

[12] G. Werner-Allen, K. Lorinez, M. Welsh, O. Marcillo, J. Jonson, M. Ruiz, and J. Lees, Deploying a Wireless Sensor Nnetwork on an Active Volcano, *IEEE Internet Computing* **10**(2) (2006), 18–25.

[13] M. R. Yuce, S. W. Ng, N. L. Myo, J. Y. Khan, and W. Liu, Wireless Body Sensor Network Using Medical Implant Band, *Medical Systems* **31**(6) (2007), 467–474.

[14] E. Zitzler and S. Knzli, Indicator-based Selection in Multiobjective Search, in: X. Yao, E. K. Burke, J. A. Lozano, J. Smith, J. J. Merelo Guervés, J. A. Bullinaria, J. E. Rowe, P. Tico, A. Kabán, and H.-P. Schwefel (eds.), *PPSN'04*, LNCS 3242, pp. 832–842, Springer, Berlin Heidelberg, 2004.

Author information

Stefka Fidanova, Institute of Information and Communication Technologies, Bulgarian Academy of Sciences, Sofia, Bulgaria.
Email: stefka@parallel.bas.bg

Pencho Marinov, Institute of Information and Communication Technologies Bulgarian Academy of Sciences, Sofia, Bulgaria.
Email: pencho@parallel.bas.bg

Enrique Alba, E. T. S. I. Informática, Grupo GISUM (NEO), University of Málaga, Málaga, Spain.
Email: eat@lcc.uma.es

Chapter 11

A Two-Dimensional Lorentzian Distribution for an Atomic Force Microscopy Simulator

Lado Filipovic and Siegfried Selberherr

Abstract. Atomic force microscopy (AFM) is a lithographic technique capable of manufacturing nanometer-sized devices. A Monte Carlo simulator for oxide growth on a silicon substrate using the AFM method is described. In previous publications, it is shown that a nanodot formed using AFM has a diagonal cross section that closely resembles the Lorentzian distribution. Therefore, an essential part of the simulator is the generation of particles using a two-dimensional Lorentzian distribution around the AFM needle tip. A successful algorithm was generated by integrating the probability distribution, while taking into consideration two-dimensional pieces of the surface. A second evenly-distributed value is used to generate a radial distribution θ. The location of each generated particle is then described by $(x,y) = (r_x \cos\theta, r_x \sin\theta)$.

Keywords. Atomic Force Microscopy, AFM, Two-Dimensional Lorentzian Distribution.

Mathematics Subject Classification 2010. 00A72, 65C05, 65C10, 65C17.

11.1 Introduction

Atomic force microscopy (AFM) [1] was developed in 1986 as a method to detect depressions and protuberances on a nanometer-sized section of a sample surface. Following this initial discovery, the use of an atomic force microscope was also shown to be an effective tool in the oxidation of nanometer-sized patterns on a silicon wafer [2]. Furthermore, it has been extensively used not only in the semiconductor industry, but also in physics, chemistry, biology, biochemistry, and other disciplines where the chemical or physical properties of a surface are required [3]. The AFM lithographic technique is capable of manufacturing nanometer-sized devices and is a promising alternative to modern lithographic methods. Performing physical simulations to replicate the AFM process is currently not feasible for large surface simulations, and therefore a Monte Carlo approach must be considered. Previous research has shown that a simulator for the generation of AFM nanodots is feasible in a Level Set environment [4] using empirical equations for the nanodot height and full width at half maximum (FWHM). It has also been shown that the physical shape of an oxide dot, generated using AFM, most closely resembles a Lorentzian distribution when viewed in diagonal cross sections [5]. Therefore, it is the goal of this work to create a Monte Carlo simulator for an AFM nanodot following a two-dimensional Lorentzian distribution around the AFM needle tip.

The simulator must be seamlessly integrated into an existing Level Set simulator environment, which is described in detail in [6]. The wafer surface and surfaces where differing materials intercept, such as the silicon–silicon-dioxide interface, are described using the Level Set equation. The equation describing the effect of the processing conditions, including bias voltage, process time, and ambient humidity, on the height and FWHM of the oxide nanodot is

$$H(t,V,h) = [H_1(V) + H_2(V)\,ln(t)]\left[0.00037h^2 - 0.019h + 0.928\right],$$
$$W(t,V,h) = [(11.6 + 9V) + (2.7 + 0.9V)\ln(t)]\,[0.019h - 0.051],$$

where $H_1 = -2.1 + 0.5V - 0.006V^2$ and $H_2 = 0.1 + 0.03V - 0.0005V^2$.

11.2 Modeling Oxidation Kinetics

Figure 11.1a shows a simplified version of the kinetics of AFM oxidation. A charged AFM tip is brought near a grounded silicon surface, resulting in the generation of an electric field. This field breaks up the ambient into ions and causes the downward acceleration of oxyions (O^-, OH^-) towards the surface. The combination of these ions with the silicon surface results in the generation of SiO_2. The model implemented to mimic oxidation kinetics is similar to the method described in [5]. Particles are generated around the needle tip with a desired two-dimensional distribution. Each particle is

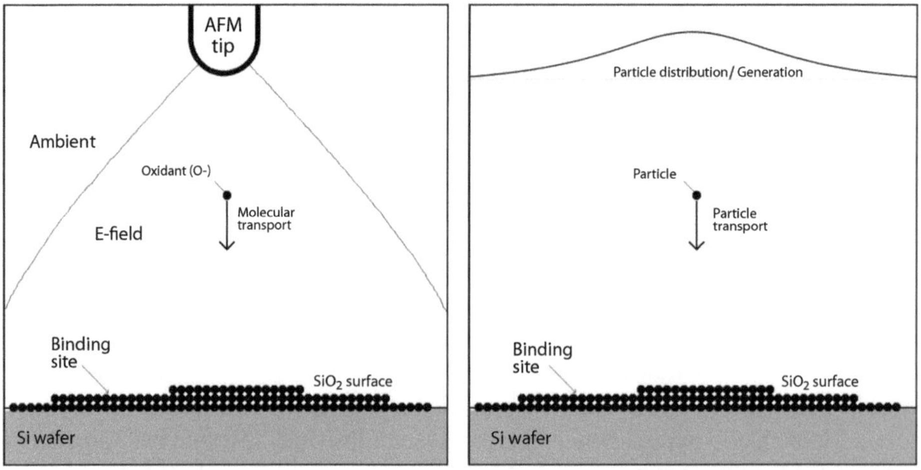

(a) Oxidation kinetics. (b) Oxidation kinetics model.

Figure 11.1. The AFM tip is modeled by generating particles generated around the needle tip, following a desired distribution.

accelerated towards the silicon surface, causing a "bump" upon impact. The summation of all particles generates the desired nanodot shape on the surface. In [5] particles are generated using an even distribution. The Monte Carlo rejection technique is then implemented to select the particles to be used for the nanostructure. The Monte Carlo rejection techniques is not feasible when nanosized structures are required to be simulated on millimeter sized wafers.

Therefore, a more direct Monte Carlo approach must be taken to simulate nanodots and nanowires. The implementation of the simulator is as follows:

1. Generate a particle at position $p_0(x_0,y_0,z_0)$, located on a plane parallel to the silicon surface with a desired distribution (Gaussian or Lorentzian), as shown in Figure 11.1b. x_0 and y_0 are distributed random variables, while $z_0 = d$ is the effective vertical position of the static dot charge.
2. Accelerate the particle towards the silicon surface along the vertical direction, until it collides with the top surface.
3. At the impact location, advance the ambient–oxide interface towards the ambient while advancing the oxide–silicon interface into the silicon.
4. If the number of particles is 0 the simulation is complete. Otherwise particle counter must be reduced by 1 and the procedure must be repeated from Step 1.

The processing steps required to simulate AFM using a Monte Carlo method with a Gaussian or Lorentzian distribution of particles is shown in Figure 11.2.

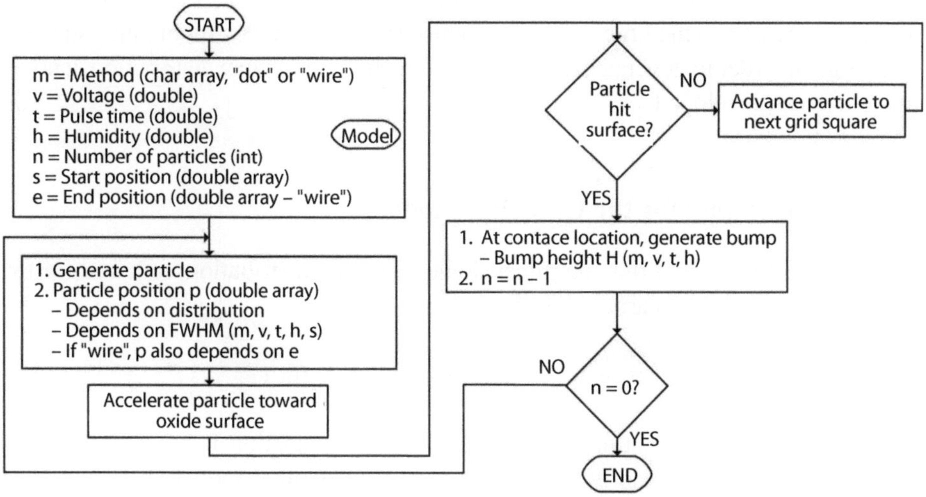

Figure 11.2. Flow chart for the Monte Carlo AFM simulator in a Level Set environment.

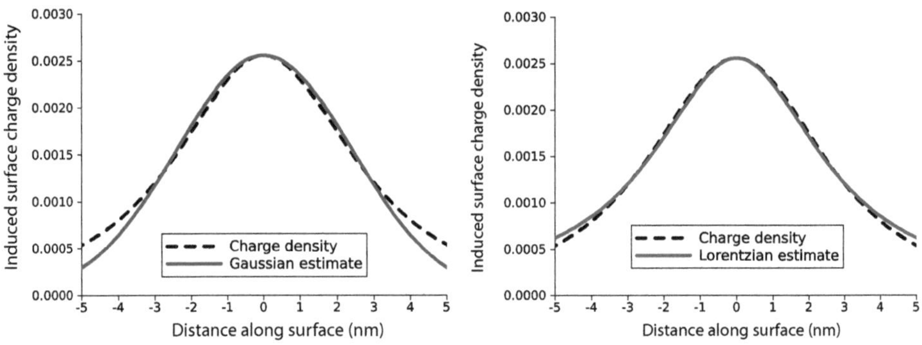

(a) Gaussian distribution. (b) Lorentzian distribution.

Figure 11.3. Difference between Gaussian and Lorentzian distributions.

11.3 Development of the Lorentzian Model

As seen from the previous discussion regarding the Monte Carlo model for AFM oxidation from Figure 11.2, a method to distribute particles according to a desired distribution is required. The Gaussian distribution is well known; however, in a previous publication [5] it was found that a Lorentzian distribution is a better fit to the final shape of a desired nanostructure, as can be seen in Figure 11.3.

The implementation of the Gaussian distribution was performed successfully, while a similar approach to the Lorentzian distribution was attempted without much success. Therefore, in order to generate particles according to a Lorentzian distribution, a novel approach was developed.

11.3.1 Algorithm for the Gaussian Model

The quantile function of the one-dimensional Gaussian distribution, required for the generation of a random particle position x_r, is

$$x_r \equiv \sqrt{2}\,\mathrm{erf}^{-1}(2u-1) \qquad u \in (0,1)\,. \tag{11.1}$$

Because of the error function, the quantile Gaussian function is not easily implementable in a Monte Carlo environment, and hence another model is desired. The model implemented is based on the well-known Marsaglia polar method [7]. This method suggests a way to generate two independent standard normal random variables. The first step is the generation of an evenly distributed random point (r_x,r_y) within a circle of unity radius $s = r_x^2 + r_y^2$, where r_x and r_y are evenly distributed random numbers $\in (-1, 1)$. The Gaussian distributed coordinates (x,y) can then be

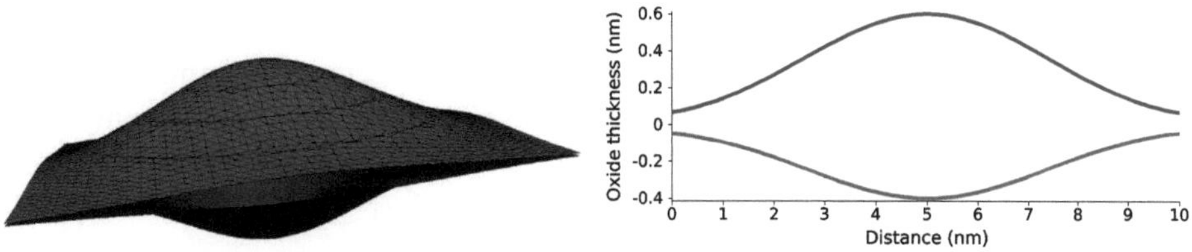

(a) Gaussian nanodot and its (b) diagonal cross section.

Figure 11.4. Nanodot generated using Gaussian particle distribution.

calculated using the Marsaglia equations:

$$x = r_x \sqrt{\frac{-2\ln(s)}{s}}, \qquad y = r_y \sqrt{\frac{-2\ln(s)}{s}}. \tag{11.2}$$

A sample Gaussian distributed nanodot is shown in Figure 11.4.

11.3.2 Development of the Lorentzian Model

11.3.2.1 One-Dimensional Lorentzian Distribution

The normalized pdf (probability density function) of the Lorentzian distribution is given by

$$f(x) = \frac{1}{\pi} \cdot \frac{1}{1+x^2}. \tag{11.3}$$

The cpd (cumulative probability distribution) is found by integrating the probability density function $\Phi(x) = \int_{-\infty}^{x} f(u)\,du$

$$\Phi(x) = \frac{1}{\pi} \arctan(x) + \frac{1}{2}. \tag{11.4}$$

The quantile function of the Lorentzian distribution, required for particle generation, is the inverse cpd

$$x_r = \Phi^{-1}(u) \equiv \tan\left[\pi\left(u - \frac{1}{2}\right)\right], \tag{11.5}$$

where $u \in (0, 1)$ is a uniformly distributed random number.

11.3.2.2 Two-Dimensional Lorentzian Distribution

We must perform the same analysis shown for the one-dimensional Lorentzian distribution in order to generate a two-dimensional Lorentzian quantile function. The pdf of the two-dimensional Lorentzian distribution can be represented as

$$f(x, y) = \frac{C}{1+x^2+y^2}, \tag{11.6}$$

where C is the normalization constant. Using polar coordinates, where $x^2 + y^2 = r^2$ and $dx\,dy = r\,dr\,d\theta$, it can easily be shown that the pdf cannot be normalized in the whole $\Re^2$. We therefore must normalize the equation to a desired maximum radius $r_{\max}$:

$$1 = 2\pi\,C \int_0^{r_{\max}} \frac{1}{1+r^2} r\,dr = C\,\pi\,\ln\left(1 + r_{\max}^2\right). \tag{11.7}$$

The normalization constant follows from the cpd, normalized to $r_{\max}$:

$$\Phi\left(r\right) = \frac{\ln\left(1 + r_L^2\right)}{\ln\left(1 + r_{\max}^2\right)}, \tag{11.8}$$

where we can treat $\ln\left(1 + r_{\max}^2\right)$ as a $r_{\max}$-dependent constant M. By inverting the cpd and solving for r_L we can find the two-dimensional Lorentzian quantile function, required for particle generation,

$$r_L = \sqrt{e^{u\,M} - 1}, \tag{11.9}$$

where $u \in(0,1)$ is a uniformly distributed random number.

An evenly-distributed angle θ between 0 and 2π is generated, and the final particle position is given by (x,y) = (r_x cos θ,r_x sin θ). It can be observed that the choice of $r_{\max}$ affects the height of the Lorentzian distribution, thereby affecting the height of the desired nanodot. Therefore, an additional contribution, dependent on $r_{\max}$ is needed in the equation for the height generated by each particle. This contribution is $\ln(1 + r_{\max}^2)$. The resulting nanodot cross section, shown in Figure 11.5 matches the ideal one-dimensional Lorentzian distribution. This method allows for the generation of nanodots, such as the one shown in Figure 11.6, which follow a Lorentzian distribution, as required for the AFM simulator.

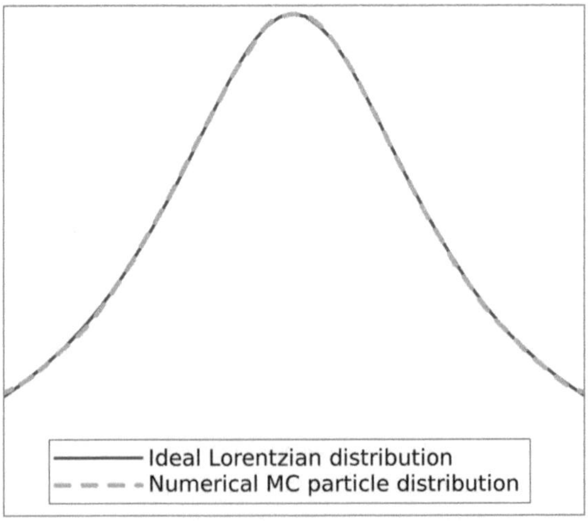

Figure 11.5. Nanodot cross section generated using a Lorentzian distribution.

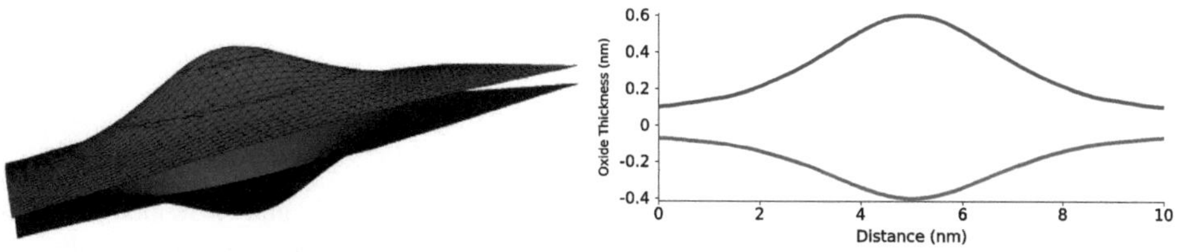

(a) Lorentzian nanodot and its (b) diagonal cross section.

Figure 11.6. Nanodot generated using Lorentzian particle distribution.

11.3.2.3 Marsaglia Polar Method

The expression for the quantile function in (11.9) suggests a potential connection to the Marsaglia polar method. A Lorentzian distribution can be generated using a similar procedure. The first step is, once again, the generation of a random point (r_x,r_y) within a unit circle with radius $s = r_x^2 + r_y^2$. The Lorentzian distributed coordinates (x,y) can then be expressed as

$$x = r_x \sqrt{\frac{e^{s\,M} - 1}{s}}, \qquad y = r_y \sqrt{\frac{e^{s\,M} - 1}{s}}, \tag{11.10}$$

where $M = \ln(1 + r_{\max}^2)$.

11.4 Conclusion

A two-dimensional Lorentzian distribution of particles is developed for an AFM simulator. The simulator implements a Monte Carlo method to generate particles around the needle tip, which are subsequently accelerated towards the wafer surface, described using the Level Set equation. Once the particles have been generated and accelerated towards the surface, the resulting structure forms an AFM nanodot, the cross section of which follows a Lorentzian probability distribution. A method, similar to the Marsaglia polar method for Gaussian particle generation is implemented to generate Lorentzian distributed particles in two-dimensional space.

Acknowledgments. We would like to thank Dr. Mihail Nedjalkov and Dr. Hans Kosina for their support in the development of this work.

References

[1] G. Binning, C. F. Quate, and C. Gerber, Atomic Force Microscopy, *Physical Review Letters* **56** (1986), 930–933.

[2] J. A. Dagata, J. Schneir, H. H. Harary, C. J. Evans, M. T. Postek, and J. Bennett, Modification of Hydrogen-Passivated Silicon by a Scanning Tunneling Microscope Operating in Air, *Applied Physics Letters* **56**(20) (1990), 2001–2003.

[3] Q. Tang, S.-Q. Shi, and L. Zhou, Nanofabrication with Atomic Force Microscopy, *Journal of Nanoscience and Nanotechnology* **4**(8) (2004), 948–963.

[4] L. Filipovic, H. Ceric, J. Cervenka, and S. Selberherr, A Simulator for Local Anodic Oxidation of Silicon Surfaces, in: *Proceedings of the 24th Canadian Conference on Electrical and Computer Engineering (CCECE) 2011*, pp. 695–698, 8–11 May 2011.

[5] L. Filipovic and S. Selberherr, A Monte Carlo Simulator for Non-Contact Mode Atomic Force Microscopy, in: I. Lirkov, S. Margenov, J. Wasnievski (eds.), *Proc. International Conference on Large-Scale Scientific Computations (LSSC), June 2011*, pp. 447–454, Lecture Notes in Computer Science 7116, Springer-Verlag, 2012.

[6] O. Ertl and S. Selberherr, A Fast Level Set Framework for Large Three-Dimensional Topography Simulations, *Computer Physics Communications* **180**(8) (2009), 1242–1250.

[7] G. Marsaglia and T. A. Bray, A Convenient Method for Generating Normal Variables, *SIAM Review* **6**(3) (1964), 260–264.

Author information

Lado Filipovic, Christian Doppler Laboratory for Reliability Issues in Microelectronics, Vienna University of Technology, Vienna, Austria.
Email: filipovic@iue.tuwien.ac.at

Siegfried Selberherr, Institute for Microelectronics, Vienna University of Technology, Vienna, Austria.
Email: selberherr@iue.tuwien.ac.at

Monte Carlo Methods and Applications, 105–113

Chapter 12
Stratified Monte Carlo Integration

Rami El Haddad, Rana Fakhreddine, and Christian Lécot

Abstract. We analyze a Monte Carlo method using stratified sampling for approximate integration. We focus on integration of non-smooth functions: we consider the indicator function of a Jordan-measurable subset of the s-dimensional unit cube $I^s := [0,1)^s$. We prove a bound for the variance and show an improved convergence rate (compared to plain Monte Carlo). When the boundary of the subset is defined by a function on I^{s-1}, the variance is estimated by means of the variation of the function. The tightness of the previous bounds is assessed through numerical experiments in dimensions $s = 2, 3$ and 4, where we compute sample variances.

Keywords. Monte Carlo Method, Numerical Integration, Stratified Sampling.

Mathematics Subject Classification 2010. 65C05.

12.1 Introduction

Many applications require evaluation of integrals. The dimensional effect is the phenomenon whereby deterministic quadrature rules deteriorate in performance as the dimension rises. Monte Carlo (MC) methods do not suffer so much from dimensional effect. Stratified sampling is a technique for increasing the efficiency of MC methods; see, e. g., [1, 6, 7, 13]. In these works, the authors show that a well-chosen stratified estimator has a variance less than or equal to the variance of the simple sampling estimator. We analyze here the following strategy: the unit cube I^s is partitioned into $N = n^s$ intervals

$$C_k := \prod_{i=1}^{s} \left[\frac{k_i - 1}{n}, \frac{k_i}{n}\right),$$

for $k = (k_1, \ldots, k_s) \in \mathbb{N}^s$ with $1 \le k_i \le n$; in each C_k, we select a point at random. This amounts to a *randomization* of the midpoint quadrature rule.

The stratified sampling mentioned above is analyzed in [2, 3, 8] for integration of smooth functions. In contrast to the previous references, the authors are able to show an improved convergence rate for the variance (compared to crude MC). Since we have experienced that some simulation methods can be reduced to numerical integration of indicator functions of irregularly-shaped domains [4, 11, 12], we consider here the case of the integral of an indicator function: so we analyze the variance of the approximate measure of a subset of I^s. We specialize further by considering the s-dimensional area under the curve of a function $f : \overline{I}^{s-1} \to \overline{I}$. The corresponding quasi-Monte Carlo (QMC) analysis was done in [10]. Here, variance upper bounds are established and numerical experiments show that the order obtained is optimal.

This chapter is organized as follows. In Section 12.2, we analyze stratified sampling for numerical integration; variance estimates are proven and the results of numerical experiments are reported. Conclusions are drawn and future perspectives are given in Section 12.3.

12.2 Numerical Integration

For an integer $s \geq 1$, let λ_s be the s-dimensional Lebesgue measure. If $g : I^s \to \mathbb{R}$ is a square-integrable function, we want to evaluate

$$\mathcal{J} := \int_{I^s} g(x) d\lambda_s(x).$$

For the usual MC approximation, let $\{U_1, \ldots, U_N\}$ be N independent random variables uniformly distributed over I^s. Then

$$X := \frac{1}{N} \sum_{\ell=1}^{N} g \circ U_\ell$$

is an unbiased estimator of $\mathcal{J}$. Now let $\{D_1, \ldots, D_N\}$ be a partition of I^s such that

$$\lambda_s(D_1) = \cdots = \lambda_s(D_N) = \frac{1}{N}.$$

Let $\{V_1, \ldots, V_N\}$ be N independent random variables, such that, for $1 \leq \ell \leq N$, V_ℓ is uniformly distributed over D_ℓ. Then

$$Y := \frac{1}{N} \sum_{\ell=1}^{N} g \circ V_\ell$$

is another unbiased estimator of $\mathcal{J}$. This stratified sampling scheme is analyzed in [2,8]. Since Cauchy–Schwarz inequality states

$$\left(\int_{I^s} g(x) d\lambda_s(x) \right)^2 \leq N \sum_{\ell=1}^{N} \left(\int_{D_\ell} g(x) d\lambda_s(x) \right)^2,$$

we obtain $\mathrm{Var}(Y) \leq \mathrm{Var}(X)$. We specialize now to the case where g is an indicator function, which is not considered in [2, 8]. Let A be a measurable subset of I^s and $g := 1_A$. Then

$$\mathrm{Var}(X) = \frac{1}{N} \lambda_s(A)\bigl(1 - \lambda_s(A)\bigr) \leq \frac{1}{4N}.$$

We want to estimate $\operatorname{Var}(Y)$. We use the maximum norm: $\|x\|_\infty := \max_{1\le i\le s} |x_i|$. For $\varepsilon > 0$, let

$$A_\varepsilon := \{x \in I^s : \exists y \in A \ \|x-y\|_\infty < \varepsilon\},$$
$$A_{-\varepsilon} := \{x \in I^s : \forall y \in I^s \setminus A \ \|x-y\|_\infty \ge \varepsilon\}.$$

One has $A_{-\varepsilon} \subset A \subset A_\varepsilon$. We restrict ourselves to the case $N = n^s$ for some integer n, and we consider the following partition of I^s. For $k = (k_1, \dots, k_s) \in \mathbb{N}^s$ with $1 \le k_i \le n$, let

$$C_k := \prod_{i=1}^{s} \left[\frac{k_i - 1}{n}, \frac{k_i}{n}\right).$$

Proposition 12.1. *Let $A \subset I^s$ be a measurable subset of I^s and $\{V_k : 1 \le k_i \le n\}$ be N independent random variables, such that V_k is uniformly distributed over C_k. Let*

$$Y := \frac{1}{N} \sum_k 1_A \circ V_k.$$

If there exists a positive nondecreasing function γ defined on $[0, +\infty)$ such that

$$\forall \varepsilon > 0 \quad \max\left(\lambda_s(A_\varepsilon \setminus A), \lambda_s(A \setminus A_{-\varepsilon})\right) \le \gamma(\varepsilon),$$

then

$$\operatorname{Var}(Y) \le \frac{1}{2N} \gamma\left(\frac{1}{N^{1/s}}\right).$$

Proof. We have

$$\bigcup_{C_k \cap A \neq \emptyset \text{ and } C_k \not\subset A} C_k \subset A_{1/n} \setminus A_{-1/n}.$$

Hence, if $\#K$ denotes the number of elements of a set K,

$$\frac{1}{N} \#\{k : C_k \cap A \neq \emptyset \text{ and } C_k \not\subset A\} \le 2\gamma\left(\frac{1}{n}\right).$$

Since

$$\operatorname{Var}(Y) \le \frac{1}{4N^2} \#\{k : C_k \cap A \neq \emptyset \text{ and } C_k \not\subset A\}, \tag{12.1}$$

we obtain the result. □

Usually, γ is a linear function, so that the upper bound of the variance is of order $\mathcal{O}(N^{-1-1/s})$. It is noticeable that we are able to show an improved convergence rate for the variance, whereas in [1, 6, 7, 13] only variance reduction is established.

Now let us consider a function $f : \overline{I}^{s-1} \to \overline{I}$ and define

$$A_f := \{(x', x_s) \in I^s : x_s < f(x')\}.$$

We want to approximate

$$\mathcal{I} := \int_{I^{s-1}} f(x')d\lambda_{s-1}(x') = \int_{I^s} 1_{A_f}(x)d\lambda_s(x).$$

We use the classical *hit-or-miss* method, which should not be considered for integration problems [5, 9]. But some MC particle simulations may be described by the integration of an indicator function; see [4] for random walk and [11] for aerosol dynamics. We utilize the concept of variation of a function in the sense of Hardy and Krause (we refer to [15]). The following lemma is proven in [10]:

Lemma 12.2. *Let f be a function of bounded variation $V(f)$ on $\overline{I}^s$. Let $n_1, \ldots, n_s$ be integers. For $k = (k_1, \ldots, k_s) \in \mathbb{N}^s$ with $1 \le k_i \le n_i$, denote $C_k := \prod_{i=1}^{s}[(k_i - 1)/n_i, k_i/n_i)$ and $x_k, y_k \in \overline{C_k}$. Then*

$$\sum_k |f(x_k) - f(y_k)| \le V(f) \prod_{i=1}^{s} n_i \sum_{i=1}^{s} \frac{1}{n_i}.$$

Proposition 12.3. *Let $f : \overline{I}^{s-1} \to \overline{I}$ be a function of bounded variation $V(f)$ and $\{V_k : 1 \le k_i \le n\}$ be N independent random variables, such that V_k is uniformly distributed over C_k. Let*

$$Y := \frac{1}{N} \sum_k 1_{A_f} \circ V_k.$$

Then

$$\operatorname{Var}(Y) \le \left(\frac{s-1}{4} V(f) + \frac{1}{2} \right) \frac{1}{N^{1+1/s}}.$$

Proof. For $k = (k_1, \ldots, k_s) \in \mathbb{N}^s$ with $1 \le k_i \le n$, we put $k' = (k_1, \ldots, k_{s-1})$ and $C'_{k'} := \prod_{i=1}^{s-1}[(k_i - 1)/n, k_i/n)$. We start from (12.1):

$$\operatorname{Var}(Y) \le \frac{1}{4N^2} \sum_{k'} \#\{k_s : C_{(k',k_s)} \cap A_f \ne \emptyset \text{ and } C_{(k',k_s)} \not\subset A_f\}.$$

We have the following inferences:

- if $C_{(k',k_s)} \cap A_f \ne \emptyset$, there exists $x'_{k'} \in C'_{k'}$ such that $k_s < nf(x'_{k'}) + 1$;
- if $C_{(k',k_s)} \not\subset A_f$, there exists $y'_{k'} \in C'_{k'}$ such that $nf(y'_{k'}) < k_s$.

Hence

$$\operatorname{Var}(Y) \le \frac{1}{4N^2} \sum_{k'} \left(n \left| f(x'_{k'}) - f(y'_{k'}) \right| + 2 \right).$$

The result follows then from Lemma 12.2. □

Our variance bound clearly represents a gain in accuracy of the factor $N^{-1/s}$ as opposed to simple MC. Of course this is of diminishing importance as s becomes large and limits the use of a rectangular grid for the stratification to problems of moderate dimension. This is precisely the case in some MC particle simulations that can be described by an integration in the dimensional range $2 \le s \le 6$; see, e. g., [4, 11, 14]. One must be aware that the integrands arising in the simulation context sometimes lend themselves to a very natural way of stratifying the domain, which typically amounts to using cells obtained by partitioning the unit cube along only a few dimensions. Although getting the same kind of bound on the variance as here can be more difficult, the resulting estimator may very well have a smaller variance than our method on medium- to high-dimensional problems. Our approach is perhaps more universal, or less problem-dependent.

We use a simple example to illustrate the previous analysis. We consider the subset of the unit ball

$$Q := \{x \in I^s : \|x\|_2 < 1\},$$

where we denote by $\|x\|_2$ the Euclidean norm of x. We have

$$\lambda_s(Q) = \frac{\pi^{s/2}}{2^s \Gamma(\frac{s}{2} + 1)}.$$

We approximate this measure and we compare the errors produced by the usual MC method, the stratified sampling MC scheme described above, and the QMC method using a Faure sequence [15]. The results are displayed in Figure 12.1. Stratified strategy clearly outperforms simple MC and behaves like QMC. If we assume error $= \mathcal{O}(N^{-\alpha})$, we can use linear regression to evaluate α; the outcomes are listed in Table 12.1. This exponent strongly depends on the series of pseudorandom numbers used.

In order to estimate the variance of the MC and stratified MC approximations, we can replicate the quadrature independently M times and compute the sample variance. We use $M = 100, 200, \dots, 1000$ and we only see small differences between the estimates. The results for $M = 100$ and $M = 1000$ are displayed in Figure 12.2. It is clear that the better accuracy due to stratified sampling goes beyond an improved convergence rate, at least in this very simple experiment. Assuming Var $= \mathcal{O}(N^{-\beta})$, linear regression can be used to evaluate β and the outputs (for $M = 1000$) are listed in Table 12.2. The values obtained are very close to the orders of the bounds given in Propositions 12.1 and 12.3.

Table 12.1. Order α of the error.

Dimension	*MC*	*Stratified MC*	*QMC*
2	0.44	0.78	0.86
3	0.57	0.57	0.80
4	0.63	0.69	0.68

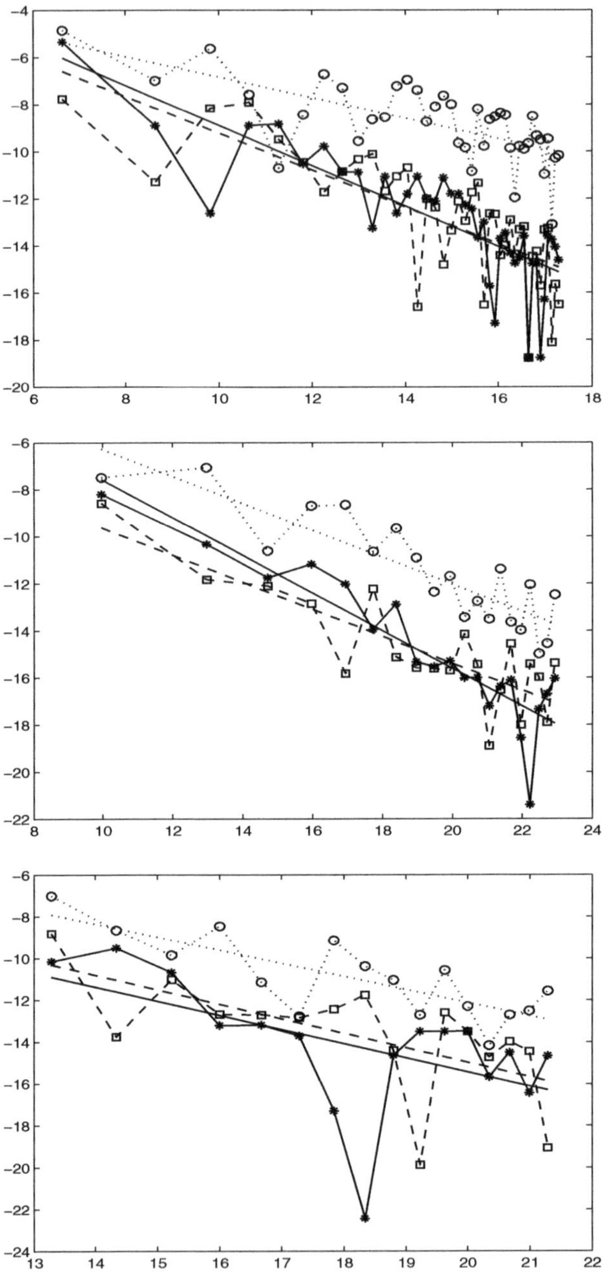

Figure 12.1. Error in the calculation of $\lambda_s(Q)$ and linear regression estimates of the error as a function of N; $s = 2$ and $10^2 \leq N \leq 400^2$ (top), $s = 3$ and $10^3 \leq N \leq 200^3$ (center), $s = 4$ and $10^4 \leq N \leq 40^4$ (bottom). Log-log plots of MC (dotted lines), stratified MC (dashed lines) and QMC (solid lines) outputs.

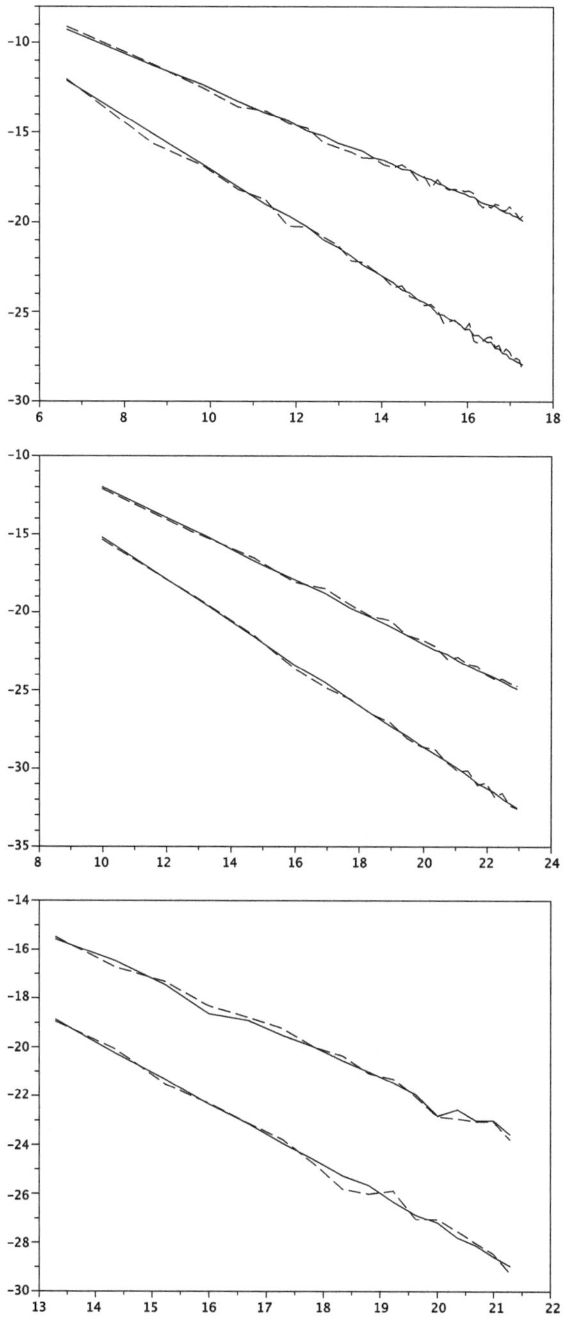

Figure 12.2. Sample variance of M independent copies of the calculation of $\lambda_s(Q)$ as a function of N; $s = 2$ and $10^2 \leq N \leq 400^2$ (top), $s = 3$ and $10^3 \leq N \leq 200^3$ (center), $s = 4$ and $10^4 \leq N \leq 40^4$ (bottom). Log-log plots of MC (top curves) and stratified MC (bottom curves) outputs; dashed lines correspond to $M = 100$ and solid lines correspond to $M = 1000$.

Table 12.2. Order β of the variance.

Dimension	*MC*	*Stratified MC*	*Theoretical bound*
2	0.99	1.49	1.50
3	1.00	1.33	1.33
4	1.00	1.25	1.25

12.3 Conclusion

A stratified sampling method was analyzed. Variance bounds were proven for integration of indicator functions. The results of numerical experiments showed that the bounds are tight. The choice of a rectangular grid for the stratification tends to restrict the use of the method to low-dimensional problems, although the potentiality of the present approach is not necessarily limited to an improvement in the convergence rate for the variance. A possibility for further work would be to use the stratified sampling strategy in simulation algorithms, as done in [12] with randomized QMC method.

Acknowledgments. We thank the reviewers for their constructive criticism on the first version of this paper.

References

[1] P. Bratley, B. L. Fox, and L. E. Schrage, *A Guide to Simulation*, 2nd ed., Springer, New York, 1987.

[2] R. C. H. Cheng and T. Davenport, The problem of dimensionality in stratified sampling, *Management Sci.* **35** (1989), 1278–1296.

[3] P. J. Davis and P. Rabinowitz, *Methods of Numerical Integration*, 2nd ed., Academic Press, San Diego 1984.

[4] R. El Haddad, C. Lécot, and G. Venkiteswaran, Diffusion in a nonhomogeneous medium: quasi-random walk on a lattice, *Monte Carlo Methods Appl.* **16** (2010), 211–230.

[5] M. Evans and T. Swartz, *Approximating Integrals via Monte Carlo and Deterministic Methods*, Oxford Statistical Science Series 20, Oxford University Press, New York, 2000.

[6] G. S. Fishman, *Monte Carlo. Concepts, Algorithms, and Applications*, Springer Series in Operations Research, Springer, New York, 1996.

[7] P. Glasserman, *Monte Carlo Methods in Financial Engineering*, Applications of Mathematics 53, Springer, New York, 2004.

[8] S. Haber, A modified Monte-Carlo quadrature, *Math. Comput.* **20** (1966), 361–368.

[9] J. M. Hammersley and D. C. Handscomb, *Monte Carlo Methods*, Methuen, London, 1964.

[10] C. Lécot, Error bounds for quasi-Monte Carlo integration with nets, *Math. Comput.* **65** (1996), 179–187.

[11] C. Lécot and A. Tarhini, A quasi-stochastic simulation of the general dynamics equation for aerosols, *Monte Carlo Methods Appl.* **13** (2007), 369–388.

[12] P. L'Ecuyer, C. Lécot and B. Tuffin, A randomized quasi-Monte Carlo simulation method for Markov chains, *Oper. Res.* **56** (2008), 958–975.

[13] N. Madras, *Lectures on Monte Carlo Methods*, Fields Institute Monographs 16, American Mathematical Society, Providence, RI, 2002.

[14] W. J. Morokoff and R. E. Caflisch, Quasi-Monte Carlo integration, *J. Comput. Phys.* **122** (1995), 218–230.

[15] H. Niederreiter, *Random Number Generation and Quasi-Monte Carlo Methods*, CBMS-NSF Regional Conference Series in Applied Mathematics 63, Society for Industrial and Applied Mathematics, Philadelphia, PA, 1992.

Author information

Rami El Haddad, Université Saint-Joseph, Département de Mathématiques, Beyrouth, Lebanon.
Email: `rami.haddad@fs.usj.edu.lb`

Rana Fakhreddine, Laboratoire de Mathématiques, Université de Savoie, Le Bourget-du-Lac, France.
Email: `Rana.Fakhereddine@etu.univ-savoie.fr`

Christian Lécot, Laboratoire de Mathématiques, Université de Savoie, Le Bourget-du-Lac, France.
Email: `Christian.Lecot@univ-savoie.fr`

Chapter 13

Monte Carlo Simulation of Asymmetric Flow Field Flow Fractionation

Oleg Iliev, Tigran Nagapetyan, and Klaus Ritter

Abstract. SDEs are used in the description of a variety of industrial processes. While numerous studies on Monte Carlo and multilevel Monte Carlo methods for SDEs used in financial engineering (option pricing, calculation of Greeks etc.) are available in the literature, technical applications are not so broadly studied. In this paper we briefly explain the Asymmetric Flow Field Flow Fractionation process for nanoparticle separation, introduce an SDE-based mathematical model, describe a Monte Carlo procedure for its solution, and end by presenting some numerical results.

Keywords. Multilevel Monte Carlo, SDE, Nanoparticle Transport and Separation, Asymmetric Flow Field Flow Fractionation.

Mathematics Subject Classification 2010. 65C05.

13.1 Motivation

The motion of small particles suspended in liquids is of great importance for a number of processes in medicine, biology, chemistry, etc. Our research, in particular, is motivated by the Field Flow Fractionation approach for segregation of small particles of two or more different sizes of submicron scale (for details see, e.g., [11] and references therein).

Monte Carlo methods and their modifications are suitable tools for the numerical simulation of such processes. The multilevel Monte Carlo method [6, 8], in the following denoted by MLMC, has already proven its efficiency for many financial problems where stochastic differential equations (SDEs) are used (see, e.g., [2, 6, 7]). Furthermore, suitable MLMC methods are almost optimal randomized algorithms for the quadrature of Lipschitz functionals of SDEs [5]. We extend the application of the MLMC method to the technical problem of nanoparticle separation, as we consider an SDE-based model for a dilute suspension of rigid noninteracting particles.

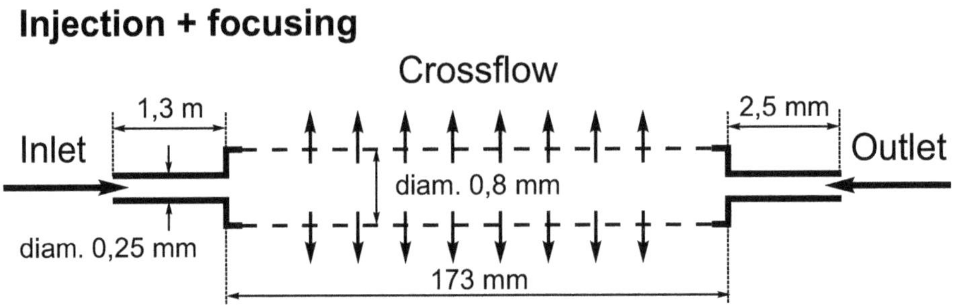

Figure 13.1. Hollow Fiber device during the focusing–injection stage for AFFFF.

13.2 AFFFF

Asymmetric Flow Field Flow Fractionation, denoted hereon as AFFFF, is a special case of the Field Flow Fractionation technique. During AFFFF the particles get separated according to their size, or more exactly, to their hydrodynamic radius. The fractionation relies on the interplay between laminar flow and Brownian diffusion. A horizontal flow of a solvent along a membrane is combined with a strong cross flow across the membrane. The membrane is impermeable for the particles, while the solvent can flow throughout it. Computer simulations of different AFFFF devices were performed by the authors, but due to the limited size of this chapter we will consider only a hollow fiber (HF) device.

AFFFF consists of two main stages: a focusing–injection stage and an elution stage.

Focusing–injection stage. An HF geometry and a sketch of flow directions during the focusing–injection stage can be seen in Figure 13.1. It should be noted that different scales for different parts of the device are used in the sketch. The axisymmetric HF, which is considered here, is 173 mm long itself (see the middle of the sketch), while the left capillary (serving to connect the HF to the sample reservoir) is 1300 mm long. During the focusing–injection stage, the solvent is entering the Hollow Fiber from both sides, and leaves through the membrane (see Figure 13.1). The ratio between the left and the right volumetric fluxes determines the position of the focusing line. The particles are injected from the left side during a certain time interval (shorter than the total duration of the focusing–injection stage), and they are transported towards the membrane due to the strong cross flow. On the other hand, the Brownian diffusion, which acts isotropically, prevents particles from resting at the membrane surface. The interplay between the force induced by the cross flow and the Brownian diffusion results in forming a boundary layer with average distance from the membrane depending on the particle size (i. e., on the diffusion coefficient) and on the intensity of the cross flow (see, e.g., [3, 13]). Smaller particles with their larger diffusion coefficient form a layer which has a larger distance to the membrane.

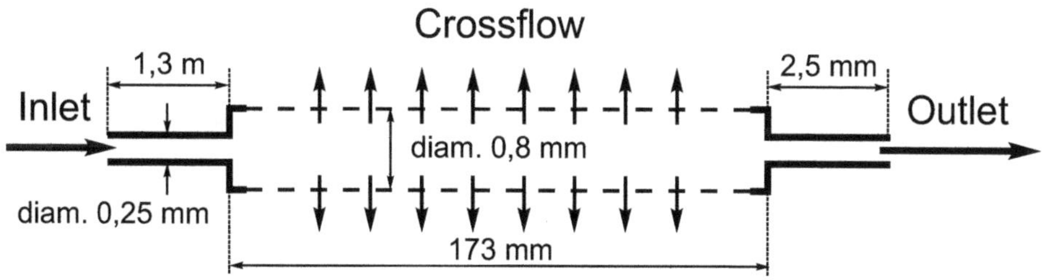

Figure 13.2. Hollow Fiber device during the elution stage for AFFFF.

At the end of the focusing–injection stage, the injected particles are located in a thin layer on the membrane, within a focusing zone around the focusing line, and this is the starting point of the consecutive horizontal transport of the particles.

Elution stage. A sketch of flow directions during the elution stage can be seen in Figure 13.2. Due to the parabolic profile of the horizontal velocity, a horizontal separation of the particles is achieved at a certain distance from the focusing line. During the elution stage, smaller particles are transported faster than the larger particles towards the channel outlet, because they experience a higher tangential flow velocity. For a more detailed description of AFFFF process see, e. g., [11, 13].

13.3 Mathematical Model and Numerical Algorithm

13.3.1 Mathematical Model

The Navier–Stokes–Brinkman system governs the flow in the channel and in the porous membrane. The equations read as follows:

$$\begin{gathered} \tfrac{\partial \vec{V}}{\partial t} - \nabla \cdot \left(\nu \nabla \vec{V}\right) + \left(\vec{V}, \nabla\right)\vec{V} + \nu K^{-1} + \tfrac{1}{\rho}\nabla p = \vec{f} \\ \nabla \cdot \vec{V} = 0. \end{gathered} \tag{13.1}$$

Here $\vec{V} = (V_1, V_2)$, p, ν and ρ are the velocity, the pressure, the viscosity and the density of the fluid, respectively (see, e. g., [1, 10]). Further on, K is the permeability of the membrane, which is considered as a porous medium here. In the pure fluid region K is set to infinity, which cancels the corresponding term, and (13.1) reduces to the Navier–Stokes system.

The particle transport in the pure fluid region $\mathcal{O} \subset \mathbb{R}^2$ is governed by the velocity field and by a stochastic term involving a 2-dimensional Brownian motion $\vec{W} = (W_1, W_2)$. The particle position $\vec{X} = (X_1, X_2)$ is a stochastic process with continuous paths taking values in $\overline{\mathcal{O}}$. This process is described by an SDE, driven by $\vec{W}$, with drift coefficient $\mu = \vec{V}$ and with a constant diffusion coefficient $\hat{\sigma} = diag(\sigma) \in \mathbb{R}^{2\times 2}$ for some diffusion parameter $\sigma > 0$. Since particles cannot penetrate the membrane, we consider all boundaries of the channel to be reflecting ones, except for the inflow and the outflow boundaries. Here the reflection is understood in the sence of normal reflection (see [12]).

Equations of this type are called multidimensional Skorohod SDEs, and existence and uniqueness results for a strong solution are available under different assumptions on the domain $\mathcal{O}$ (and for more general driving processes). In particular, the assumptions from [12] are met in our application.

13.3.2 The MLMC Algorithm

We want to estimate the expectation $E(u(\vec{X}))$ of a square-integrable functional u on the path space of the Skorohod SDE via the MLMC method based on the Euler-Maruyama scheme.

At first the drift coefficient $\mu = \vec{V}$ has to be determined numerically. To this end the Navier–Stokes–Brinkman equations (13.1) are discretized by means of a finite volume method on an orthogonal nonuniform grid, which is exponentially refined close to the membrane. The solution procedure is rather standard and hence is not discussed here. For the simulation of the SDE by means of the Euler–Maruyama scheme, the respective values of the velocity field are obtained from the approximate values on the grid via local interpolation.

In the case of the SDE *without* reflection, the Euler–Maruyama scheme with an equidistant stepsize h has the form

$$\vec{X}^h_{k+1} = \vec{X}^h_k + h \cdot \mu(kh, \vec{X}^h_k) + \sigma \cdot (\Delta^h \vec{W})_k \quad k = 0, 1, \ldots, \tag{13.2}$$

where $(\Delta^h \vec{W})_k = \vec{W}((k+1)h) - \vec{W}(kh)$ is a Brownian increment. By piecewise linear interpolation and taking into account the reflection at the boundary the Euler–Maruyama approximation is extended to a process with continuous paths taking values in $\overline{\mathcal{O}}$, which will be denoted by $\vec{X}^h$.

Obviously,

$$\vec{X}^h = \varphi^h(\Delta^h \vec{W}),$$

where $\Delta^h \vec{W} = ((\Delta^h \vec{W})_0, (\Delta^h \vec{W})_1, \ldots)$ and where φ^h is defined by (13.2) accompanied by a proper reflection algorithm (see [4]). The functionals u to be considered in the sequel depend on an initial segment of $\vec{X}$, whose length is given by an almost surely finite stopping time. Likewise the computation of $u(\vec{X}^h)$ involves only finitely many Brownian increments for every h with probability one.

We choose integers $M > 1$ and $L \geq 1$, and we employ Euler–Maruyama approximations with different time steps $h_\ell = M^{-\ell}$ for $\ell = 0, \ldots, L$. With N_ℓ denoting the number of replications at level ℓ, and with independent copies

$$\vec{W}_{0,1}, \ldots, \vec{W}_{0,N_0}, \ldots, \vec{W}_{L,1}, \ldots, \vec{W}_{L,N_L}$$

of the Brownian motion $\vec{W}$, the MLMC estimator Y for $E(u(\vec{X}))$ is given by

$$Y = \frac{1}{N_0}\sum_{i=1}^{N_0} U_{0,i} + \sum_{\ell=1}^{L}\left[\frac{1}{N_\ell}\sum_{i=1}^{N_\ell}\left(U_{\ell,i} - U'_{\ell,i}\right)\right],$$

where

$$U_{\ell,i} = u(\varphi^{h_\ell}(\Delta^{h_\ell}\vec{W}_{\ell,i}))$$

for $\ell = 0, \dots, L$ and

$$U'_{\ell,i} = u(\varphi^{h_{\ell-1}}(\Delta^{h_{\ell-1}}\vec{W}_{\ell,i}))$$

for $\ell = 1, \dots, L$. Note that $U_{\ell,i}$ and $U'_{\ell,i}$ are coupled via $\vec{W}_{\ell,i}$, while $U_{\ell,i}$ and $U'_{\ell+1,j}$ are independent.

In general, Y is a biased estimator, since $E(Y) = E(u(\varphi^{h_L}(\Delta^{h_L}\vec{W})))$. With v_0 denoting the variance of $u(\varphi^{h_0}(\Delta^{h_0}\vec{W}))$, and with v_ℓ denoting the variance of $u(\varphi^{h_\ell}(\Delta^{h_\ell}\vec{W})) - u(\varphi^{h_{\ell-1}}(\Delta^{h_{\ell-1}}\vec{W}))$ for $\ell = 1, \dots, L$, the variance of Y is given by $V(Y) = \sum_{\ell=0}^{L} v_\ell/N_\ell$. For the functionals u under consideration, the expected cost of Y is proportional to $\sum_{\ell=0}^{L} N_\ell/h_\ell$. According to [6], the minimization of $\sum_{\ell=0}^{L} N_\ell/h_\ell$ under the constraint $V(Y) \le \varepsilon^2/2$ for any given $\varepsilon > 0$ leads to

$$N_\ell = \left\lceil 2\varepsilon^{-2}\sqrt{V_\ell h_\ell}\left(\sum_{\ell=0}^{L}\sqrt{V_\ell/h_\ell}\right)\right\rceil. \tag{13.3}$$

The actual implementation of the MLMC algorithm involves bias estimates to appropriately choose the maximal level L and variance estimates to determine the number of replications per level; see (13.3). Furthermore, based on our numerical simulations experience we have chosen $M = 4$.

13.4 Numerical Results

In this section we will present some results from simulations of the focusing–injection and the elution stages of the AFFFF.

Simulation of the focusing–injection stage. The goal of the simulations, as defined by our industrial partner, is to study the dynamics of the injection of the particles, as well as the influence of the inflow control on the size and the shape of the focusing zone. Due to lack of space, only two snapshots of the injection process are shown here. Injection of smaller particles ($r = 3.2$ nm, Figure 13.3) and of larger particles ($r = 32$ nm, Figure 13.4) is simulated under the same flow conditions, and the snapshots are taken at the same time instance. Figures 13.3 and 13.4 only show half of the HF cross section, due to symmetry. The final shape of the focusing zone, which is not shown here, can be controlled by the inflow velocity.

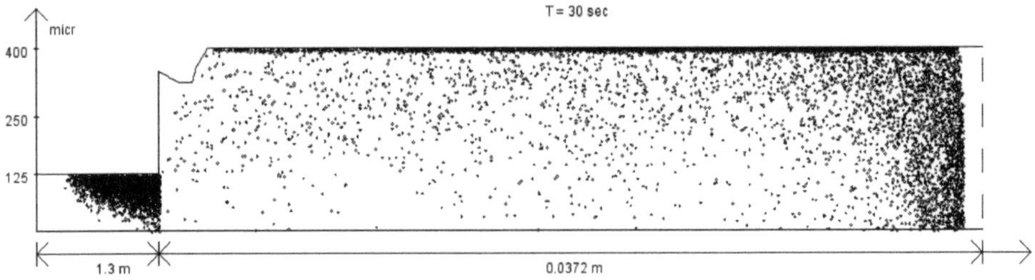

Figure 13.3. Particle injection, size $r = 3.2$ nm.

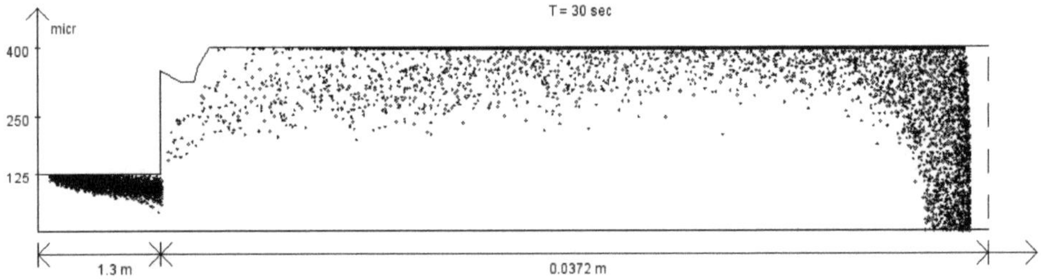

Figure 13.4. Particle injection, size $r = 32$ nm.

MLMC for restoring the diffusion coefficient at the focusing–elution stage. Let $\hat{X}_2$ denote the distance of a particle from the walls of the channel at the end of the focusing–injection stage. The cross flow $\hat{V}$ near the walls of the channel is almost a constant, so the distribution of the random variable $\hat{X}_2$ is well approximated by an exponential distribution with parameter $\hat{V}/D$, where $D = \sigma^2/2$ with σ denoting the diffusion parameter of the SDE. Recall that D depends on the particle size (see, e. g., [3]). The latter distribution is also known as a barometric distribution.

In order to check the consistency of our approach, we have employed the MLMC algorithm to obtain an empirical distribution of $\hat{X}_2$, which is then used to provide an estimate for D. The estimate is compared to the known value of D, which is calculated based on particles radius (see [3]). Results from numerical experiments with the MLMC algorithm are presented in Table 13.1.

Table 13.1. Multilevel Monte Carlo method for restoring the coefficient $D = 76.2\ \mu^2$/s. Cross flow equal to 30 μ^2/s.

Desired accuracy of MC simulation	*Absolute error*	*Relative error*
$\varepsilon = 10^{-3}$	$4.0 \cdot 10^{-1}$	$5.7 \cdot 10^{-3}$
$\varepsilon = 10^{-4}$	$9.8 \cdot 10^{-2}$	$1.4 \cdot 10^{-3}$
$\varepsilon = 10^{-5}$	$4.2 \cdot 10^{-2}$	$5.9 \cdot 10^{-4}$

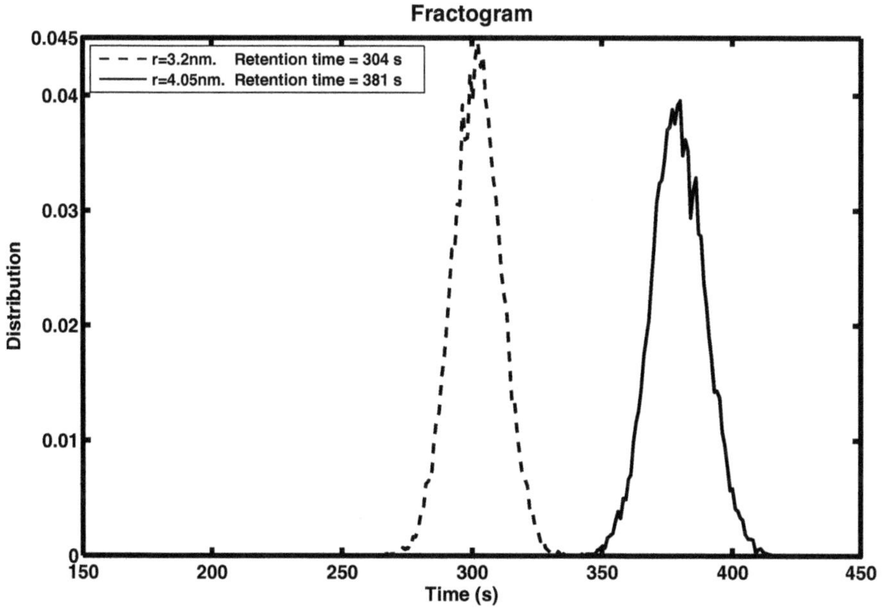

Figure 13.5. Fractogram for $r = 3.2$ nm and $r = 4.05$ nm, HF device.

Elution stage. The horizontal separation of the particles is actually done during the elution stage, and in practice the separation is evaluated from fractograms. Fractograms (see Figure 13.5) are a measure for the mass of particles that exit the outlet per unit time, and they serve as an approximation to the density of the distribution of the exit time of $\vec{X}$ from the channel. Ideally the fractograms of particles of different size should be 'well separated'. Additionally, the so-called retention times, which are the medians of the fractograms, should be 'small', as this indicates that the total duration of the separation process is small. We have employed a Monte Carlo algorithm to obtain fractograms via simulation. A particular goal of our simulations is to study the properties of the fractograms depending on different focusing and elution regimes. The results are summarized in Table 13.2. The retention time is computed for particles with $r = 3.2$ nm (fifth column) and $r = 4.05$ nm (last column) under different focusing regimes (first and third columns) and different elution regimes (fourth column). These results and the respective fractograms, which are not shown here, allow us to conclude whether or not the retention time can be reduced while keeping the fractograms well separated.

Exit times, in particular their distribution functions and quantiles thereof, play an important role in the present work. For SDEs without reflection the computation of expected exit times via classical Monte Carlo and MLMC algorithms is analyzed in [9].

Table 13.2. Parameters of simulation (absolute values). Retention times (s) of particles with radius 3.2 and 4.05 nm.

Focusing time	*Parameter*	*Focusing [ml/min]*	*Elution [ml/min]*	$t_r = 3.2$ *nm*	$t_r = 4.05$ *nm*
	inlet	0.085	1.2		
V1 — 180 sec	outlet	0.765	0.35	308	387
	crossflow	0.85	0.85		
	inlet	0.17	1.2		
V2 — 180 sec	outlet	0.68	0.35	291	366
	crossflow	0.85	0.85		
	inlet	0.34	1.2		
V3 — 180 sec	outlet	0.51	0.35	255	322
	crossflow	0.85	0.85		
	inlet	0.17	0.775		
V4 — 180 sec	outlet	0.255	0.35	160	201
	crossflow	0.425	0.425		

Acknowledgments. The authors express sincere thanks to Dr. Johann and Dr. Schuch from Wyatt Technology Europe Ltd. for providing AFFFF data and for the useful discussions on the results, and to Mrs. Shklyar from Fraunhofer ITWM for her valuable help in performing the simulations. The research of O. Iliev and T. Nagapetyan is supported by BMBF project 03MS612D, "FROPT".

References

[1] P. Angot, Analysis of singular perturbations on the Brinkman problem for fictitious domain models of viscous flows, *Math. Methods Appl. Sci.* **22** (1999), 1395–1412.

[2] R. Avikainen, On irregular functionals of SDEs and the Euler scheme, *Finance and Stochastics*, **13** (1999), 381–401.

[3] R. Becker, *Theorie der Wärme*, Springer, Heidelberg, 1966.

[4] C. Costantini, B. Pacchiarotti, and F. Sartoretto, Numerical approximation for functionals of reflecting diffusion processes, *SIAM J. Appl. Math.* **58** (1998), 73–102.

[5] J. Creutzig, S. Dereich, T. Müller-Gronbach, and K. Ritter, Infinite-dimensional quadrature and approximation of distributions, *Found. Comput. Math.* **9**, 391–429, 2009.

[6] M. Giles, Multi-level Monte Carlo path simulation, *Operations Research* **56** (2008), 607–617.

[7] M. Giles, D. Higham, and X. Mao, Analysing multi-level Monte Carlo for Options with non-globally Lipschitz payoff, *Finance and Stochastics* **13** (2009), 403–413.

[8] S. Heinrich, Monte Carlo complexity of global solution of integral equations, *J. Complexity* **14** (1998), 151–175.

[9] D. Higham, X. Mao, M. Roj, Q. Song, and G. Yin, Mean exit times and the multi-level Monte Carlo method, Research Report, University of Strathclyde, 2011.

[10] M. Kaviany, *Principles of Heat Transfer in Porous Media*, Springer, New York, 1995.

[11] M. E. Schimpf, K. D. Caldwell, and J. C. Giddings, *The FFF Handbook*, Wiley, New York, 2000.

[12] Y. Saisho, Stochastic differential equations for multi-dimensional domain with reflecting boundary, *Probab. Th. Rel. Fields* **74** (1987), 455–477.

[13] K.-G. Wahlund and J. C. Giddings, Properties of an asymmetrical flow field-flow fractionation channel having one permeable wall, *Anal. Chem.* **59** (1987), 1332–1339.

Author information

Oleg Iliev, Fraunhofer Institute for Industrial Mathematics, Kaiserslautern, Germany; Technical University of Kaiserslautern, Germany; Institute of Mathematics, Bulgarian Academy of Sciences, Sofia, Bulgaria.
Email: `oleg.iliev@itwm.fhg.de`

Tigran Nagapetyan, Fraunhofer Institute for Industrial Mathematics, Kaiserslautern, Germany; Technical University of Kaiserslautern, Germany.
Email: `tigran.nagapetyan@itwm.fhg.de`

Klaus Ritter, Technical University of Kaiserslautern, Germany.
Email: `ritter@mathematik.uni-kl.de`

Chapter 14
Convexization in Markov Chain Monte Carlo

Dimitri Kanevsky and Avishy Carmi

Abstract. Markov Chain Monte Carlo (MCMC) processes are in general governed by non-convex objective functions that are difficult to optimize. Standard regularization of MCMC processes (e. g., with quadratic penalties) in general improve optimization performance accuracy but significantly slow down optimization processes. In our paper we suggest a different convexization recursion process based on local transformations of objective functions into auxiliary functions. When auxiliary functions are created, a recursive MCMC process is then applied to these auxiliary functions to update parameters. Then thes updated parameters are used to create a new auxiliary function, and so on. We demonstrate our method on a compressive sensing problem with small dimensionality matrices.

Keywords. Markov Chain Monte Carlo Filtering, Subgradient Projection, Convex Optimization, Particle Filtering.

Mathematics Subject Classification 2010. 60G35.

14.1 Introduction

The goal of this chapter is to introduce a general convexization process for arbitrary functions to assist Markov Chain Monte Carlo (MCMC) optimization. In general there are various efficient methods to optimize convex functions. It is natural in optimization of nonconvex functions to use convex lower-bound functions in intermediate steps. Since MCMC processes in general are governed by nonconvex objective functions, this makes an optimization process based on them difficult. Standard regularization of MCMC processes , while improving optimization accuracy performance, slows them down significantly. In our chapter we suggest a different convexization recursion process which is based on a recently introduced method to transform locally any objective function into auxiliary concave functions [5]. The MCMC recursion is then applied to these auxiliary functions, and the parameter update is performed. This update process is used to create a new convex auxiliary function, and so on. We demonstrate our method on a compressive sensing problem with small dimensionality matrices, applying the Metropolis–Hastings method with auxiliary functions to a Bayesian objective function.

The rest of this chapter is structured as follows. In Section 14.2 we introduce a notion of an auxiliary function and describe how to build it. In Section 14.3 we extend this method to stochastic functions. In Section 14.4 we describe the main innovative result of this chapter: integration of auxiliary functions into the Metropolis–Hastings method. In Section 14.5 we demonstrate numerically the Metropolis–Hastings method with the auxiliary functions.

14.2 Auxiliary Functions

14.2.1 Definition of Auxiliary Functions

Let $f(x) : \mathcal{U} \subset \mathbb{R}^n \to \mathbb{R}$ be a real valued differentiable function in an open subset $\mathcal{U}$. Let $\mathbf{Q}_f = \mathbf{Q}_f(x, y) : \mathbb{R}^n \times \mathbb{R}^n \to \mathbb{R}$ be twice differentiable in $x \in \mathcal{U}$ for each $y \in \mathcal{U}$. We define $\mathbf{Q}_f$ as an auxiliary function for f in $\mathcal{U}$ if the following properties hold.

1. $\mathbf{Q}_f(x, y)$ is a strictly concave function of x for any $y \in \mathcal{U}$ with its (unique) maximum point belonging to $\mathcal{U}$ (recall that twice differentiable function is strictly concave or convex over some domain if its Hessian function is positive or negative definite in the domain, respectively).
2. Hyperplanes tangent to manifolds defined by $z = g_y(x) = \mathbf{Q}_f(x, y)$ and $z = f(x)$ at any $x = y \in \mathcal{U}$ are parallel to each other, i. e.,
$$\nabla_x \mathbf{Q}_f(x, y)|_{x=y} = \nabla_x f(x). \tag{14.1}$$
3. $f(x) = \mathbf{Q}_f(x, x)$ for any $x \in \mathcal{U}$.
4. $f(x) \geq \mathbf{Q}_f(x, y)$ for any $x, y \in \mathcal{U}$.

In an optimization process via an $\mathbb{Q}$-function, it is usually assumed that finding an optimum of an $\mathbb{Q}$-function is "easier" than finding a (local) optimum of the original function f. Naturally, a desired outcome is for the equation $\nabla_x \mathbb{Q}_f(x, y) = 0$ to have a closed form solution.

14.2.2 Optimization Process for Auxiliary Functions

The optimization recursion via an auxiliary function can be described as follows (where we use the expectation-maximization (EM) style).

E-step: given x^t construct $\mathbb{Q}_f(x, x^t)$.

M-step: find
$$x^{t+1} = \arg\max_{x \in \mathcal{U}} \mathbb{Q}_f(x, x^t). \tag{14.2}$$

For updates (14.2) we have $f(x^{t+1}) = \mathbb{Q}_f(x^{t+1}, x^{t+1}) \geq \mathbb{Q}_f(x^t, x^{t+1}) \geq \mathbb{Q}_f(x^t, x^t) = f(x^t)$. This means that iterative update rules have a "growth" property (i. e., the value of the original function increases for the new parameters values). This is illustrated in the plot in Figure 14.1.

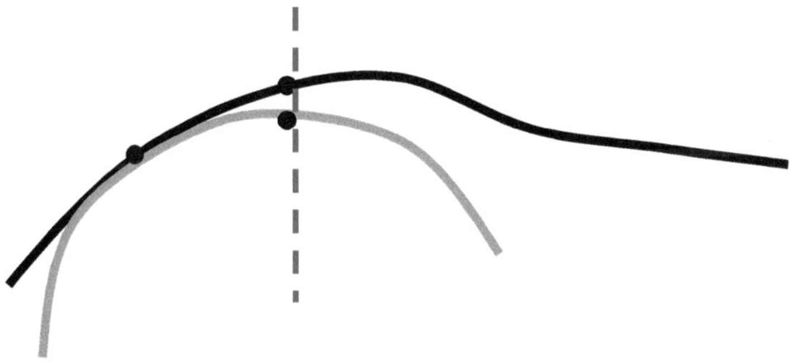

Figure 14.1. Auxiliary function.

In this figure the upper curve denotes the plot of the objective function $f : x \to \mathbb{R}$ and the curve in gray, i. e., the convex lower curve, represents an auxiliary function.

Let us call a point $x \in \mathcal{U}$ critical if $\nabla_x f(x) = 0$. We can prove the following convergence statement.

Proposition 14.1. *Let $\mathbf{Q}_f$ be an auxiliary function for f in $\mathcal{U}$ and let $\mathcal{S} = \{x^t, t = 1, 2, \ldots\}$. Then all limit points of $\mathcal{S}$ which lie in $\mathcal{U}$ are critical points. Assume in addition that f has a local maximum at some limit point of the sequence $\mathcal{S}$ in $\mathcal{U}$ and that f is strictly concave in some open neighborhood of this point. Then there exists only one critical point of $\mathcal{S}$ in $\mathcal{U}$.*

Proof. Let us define the following map:

$$T : \{x \in \mathcal{U} \to \tilde{x} = \arg\max_y \mathbf{Q}_f(y, x)\}. \tag{14.3}$$

Let us first prove that the limit points of $\mathcal{S}$ are fixed points of T. Indeed, let a be a limit point of $T^{n_i}(x), i = 1, 2, \ldots$.. Then $b = T(a)$ is a limit point for the sequence $T^{n_i+1}(x)$. If $b \neq a$ then we have

$$f(T^{n_i}(x)) \le f(T^{n_i+1}(x)) \le f(T^{n_{i+1}}(x)), \tag{14.4}$$

i. e., $F(a) \le F(b) \le F(a)$. This implies $a = b$.

Next, one can see that the fixed points of T belonging to $\mathcal{U}$ are critical points. Indeed, if $T(a) = a \in \mathcal{U}$, then $a = \arg\max_x \mathbf{Q}_f(x, a)$, and $\nabla_x \mathbf{Q}_f(x, a)|_{x=a} = \nabla_a f(a) = 0$.

The final statement of the proposition follows from the fact that if f is strictly concave in some open neighborhood of a critical point $a \in \mathcal{U}$, then for points x^t sufficiently close to a, then $x^{t+1} = \arg\max_x \mathbf{Q}_f(x, x^t)$ is close to a. □

14.2.3 Auxiliary Functions for Convex Functions

In this section, following [5], we describe how to construct auxiliary functions (which are strictly concave) for strictly convex functions.

Assume that $f(x)$ is strictly concave in $\mathcal{U}$. Then for any point $x \in \mathcal{U}$ we can construct a family of auxiliary functions as follows.

Let us consider the following family of functions:

$$\mathbf{Q}_f(y, x; \lambda) = -\lambda f\left(-\frac{y}{\lambda} + x\left(1 + \frac{1}{\lambda}\right)\right) + f(x) + \lambda f(x). \tag{14.5}$$

These family functions (14.5) obey properties 1–3 for any $\lambda > 0$ in the definition of auxiliary function. The family of functions (14.5) are obtained via subsequent applications of the following three transformations.

1. Reflection along x-axis:

$$\mathbf{H}_f(y, x) = -f(y) + 2f(x). \tag{14.6}$$

2. Reflection along y-axis:

$$\mathbf{G}_f(y, x) = \mathbf{H}_f(-y + 2x, x). \tag{14.7}$$

3. Scaling:

$$\mathbf{Q}_f(y, x; \lambda) = \lambda \mathbf{G}_f\left(\frac{y}{\lambda} + x\left(1 - \frac{1}{\lambda}\right), x\right) + (1 - \lambda)\mathbf{G}_f(x, x). \tag{14.8}$$

Various properties of these transformation are given in [5]. Specifically, one can easily see that when a scaling factor λ grows, the function $\mathbf{Q}_f(y, x; \lambda)$ becomes larger and its extremum is moving further from x. Another useful fact is that if $f(x)$ is convex, then for any $\lambda > 0, x, y \in \mathcal{U}$ one has $\mathbf{Q}_f(y, x; \lambda) \leq f(x)$.

In general, for an arbitrary function $f(x)$ one can construct auxiliary functions $\mathbf{Q}_f(y, x; \lambda)$ locally (with different λ in neighborhoods for different points x).

This method of building auxiliary functions can be extended to a more general class of functions, as shown in the next sections.

14.2.4 Objective Function Which Is the Sum of Convex and Concave Functions

Assume that

$$f(x) = g(x) + h(x), \tag{14.9}$$

where $h(x)$ is strictly convex in $\mathcal{U}$. Then we can define an auxiliary function for $f(x)$ as

$$Q_f(y, x) = Q_g(y, x) + Q_h(y, x, \lambda), \tag{14.10}$$

where $Q_g(y, x)$ is some auxiliary function associated with g (e. g., coinciding with $g(x)$ if $g(x)$ is strictly concave). In practical applications some function $Q_h(y, x, \lambda)$ may be concave, but not strictly concave. In this case one can add a small regularized penalty to make it strictly concave.

14.2.4.1 Exponential Families

An important example of convex functions is the exponential family. We define an exponential family as any family of densities on $\mathbb{R}^D$, parameterized by θ, which can be written as $\xi(x, \theta) = \frac{\exp\{\theta^T \phi(x)\}}{Z(\theta)}$, where x is a D-dimensional base observation. The function $\phi : \mathbb{R}^D \to \mathbb{R}^d$ characterizes the exponential family. $Z(\theta) = \int_\Xi \exp\{\theta^T \phi(x)\} dx$ is the partition function that provides the normalization necessary for $\xi(x, \theta)$. The function $\log \xi(x, \theta)$ is convex, and it is strictly convex if $Var[\phi(x)] \neq 0$ [4]. Some objective functions of exponential densities (e. g., in energy-based models) can be optimized via a recursion procedure that at each recursion require optimization of the weighed sum of exponential densities, i. e., a sum of convex and concave functions.

14.3 Stochastic Auxiliary Functions

14.3.1 Stochastic Convex Learning (Summary)

Let us consider a convex minimization problem as follows. We assume that our goal is to find some parameter vector $x \in \mathcal{U}$ such that the sum of convex functions $f^i \to \mathbb{R}$ takes on the smallest possible value. The order of the functions f^i can be chosen in response to our previous choice of x, or the functions f^i can be drawn from some distribution. It is our goal to find a sequence of x^i such that the cumulative value (empirical loss or risk in machine learning terminology) of $f^i(x^i)$ is minimized. Let the average cumulative loss is defined as

$$f^*(x) = \frac{1}{T} \sum_{t=1}^{T} f^t(x) \tag{14.11}$$

and

$$x^* = \arg\min_{x \in \mathcal{U}} f^*(x). \tag{14.12}$$

Minimizing the function (14.11) can be done using the batch gradient descent:

$$x^{t+1} = x^t - \gamma_t \nabla_x f^*(x^t) = x^t - \gamma_t \frac{1}{T} \sum_{i=1}^{T} \nabla_x f^i(x^t). \tag{14.13}$$

The properties of this algorithm are well known (see, e. g., [1]). When learning rates of γ_t are small enough, the algorithm converges to a local minimum of $f^*(x)$. Each iteration of batch gradient descent, however, involves a burdening computation of the average of the gradient of the function $f^*(x)$ over an entire training set. Significant resources must be allocated in order to store the sufficiently large training set and compute this average. The elementary online gradient descent algorithm is obtained by dropping the averaging operation in the batch descent algorithm. Instead of averaging the gradient of the function f^* over the complete training set, each operation of the online gradient descent consists of choosing a function f^t at random (as corresponding to a random training example) and updating the parameter x^t according to the formula

$$x^{t+1} = x^t - \gamma_t \nabla_x f^t(x^t). \tag{14.14}$$

14.3.2 Auxiliary Stochastic Functions

Assume now that functions $f^i(x)$ are non-concave and we need to solve the maximization problem

$$\max \sum f^i(x). \tag{14.15}$$

Assume also that $Q^i(y,x)$ are auxiliary functions for $f^i(y)$ at x. In this case one can consider the following optimization process. Let

$$Q^*(y,x) = \sum Q^i(y,x). \tag{14.16}$$

Then $Q^*(y,x)$ is an auxiliary function for $f^*(y) = \sum f^i(y)$. For $t = 1, 2, \ldots$ we can optimize $Q^*(x^t, y)$ using stochastic descent methods and find x^{t+1}. This induces the optimization process for $f^*(x)$ via the auxiliary function $Q^*(y,x)$.

14.4 Metropolis–Hastings Auxiliary Algorithm

We will now describe how to combine the convexization process with some MCMC technique such as Metropolis–Hastings. Assume that we want to draw samples from a probability distribution $P(x)$ which is proportional to some complex (not convex) expression $f(x)$. Assume that we have an ergodic and balanced Markov chain x^t that at sufficiently long times generate states which obey the $P(x)$ distribution. Let $Q(x';x^t)$ be the proposal densities which depend on the current state x^t to generate a new proposed state x'. The proposal is "accepted" as the next value $x^{t+1} = x'$ if α is drawn from $U(0,1)$, the uniform distribution satisfies

$$\alpha < \frac{f(x')}{f(x^t)} \frac{Q(x';x^t)}{Q(x^t;x')}. \tag{14.17}$$

In our novel approach we define proposals as auxiliary functions in the Metropolis–Hastings algorithm as described below. Let $Q_f(x,y)$ be an auxiliary function for $f(x)$. Then we have the following:

- given the most recent sampled value x^t, draw a new proposal state x' with the probability $Q_f(x'; x^t)$;
- calculate
$$a = \frac{f(x')}{f(x^t)} \frac{Q_f(x'; x^t;)}{Q_f(x^t; x')}; \tag{14.18}$$
- the new state x^{t+1} is chosen according to the following rules: if $a \geq 1$ then $x^{t+1} = x'$ else $x^{t+1} = x'$ with probability a and $x^{t+1} = x^t$ with probability $1 -$.

14.5 Numerical Experiments

In the previous section we demonstrated concepts of auxiliary proposal in MCMC by considering the following optimization problem, which is a Bayesian representation of a compressive sensing problem [6]:

$$\max_x \exp\left(\frac{-0.5||y - Hx||^2}{R}\right) * \exp\left(\frac{-0.5||x||_1^2}{\sigma^2}\right). \tag{14.19}$$

In this formula y is an m dimensional vector (measurement), H is an $m \times n$ sensing matrix with $m < n$, x is an n dimensional parameter vector, and the function $\exp(\frac{-0.5||x||_1^2}{\sigma^2})$ is a "semi-Gaussian" penalty to enforce the sparsity (here $||x||_1^2 := (\sum_i |x_i|)^2$ for all entries x_i in x). We prefer to use the semi-Gaussian penalty (which uses a quadratic term $|x|_1^2$ in the exponential function) instead of a Laplacian penalty (which uses a linear term $|x|_1$ in the exponential function), since it allows as to introduce a concave function at x_0,

$$Q(x, x_0) = \lambda \exp\left(\frac{-0.5(sign(x_0) * (\frac{x}{\lambda} + (1 - \frac{1}{\lambda})x_0)^2}{\sigma^2}\right), \tag{14.20}$$

which, as shown in [7], satisfies condition 2 in Section 2.1. Therefore a lower bound function can be obtained for sufficiently small λ.

We run simulation comparative experiments using the standard Metropolis–Hastings method (14.17) and the Metropolis–Hastings method (14.18) with the convex auxiliary function for the problem (14.19). In our simulation experiments, parameters were chosen as the follows: $n = 10, m = 5$. Entries in the sensing matrix H were obtained by sampling according to $\mathcal{N}(0, 1/5)$. In our experiments the signal support vector $x \in \mathbb{R}^{10}$ is assumed to be a sparse parametric vector with signal support consisting of two elements. Note that in a usual compressive sensing approach, the dimensions of the sensing matrix obtained via random sampling are made sufficiently large in order to ensure that these matrices obey the Restricted Isometry Property [2], which guarantees with overwhelming probability the accurate recovering of a sparse parameter vector x. Since we are considering relatively small dimensional matrices,

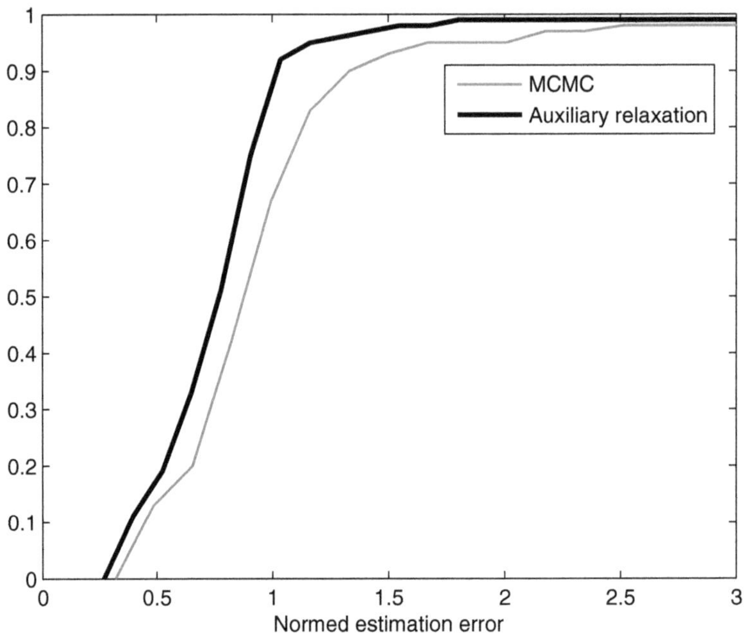

Figure 14.2. Cumulative distribution of errors.

we can no longer guarantee the accurate recovery of a "true" sparse vector, and therefore our estimation error for this problem is relatively high in comparison with large dimensional compressive sensing problems [3]. Figure 14.2 is based on 100 runs and gives the cumulative distribution of errors. In each run we produced 10,000 samples and had 5000 burn-in samples. The ordinate axis is the probability, and the absica is the normed estimation error. Every point in the figure shows the probability that the error would be smaller than the corresponding x value.

14.6 Conclusion

In this chapter we introduced a novel convexization approach for MCMC based on general convexization techniques from [5], which allow us to build auxilary functions for a wide range of problems. We illustrated this convexization method on a compressive sensing problem represented in a Bayesian form with a semi-Gaussian penalty. Simulation experiments showed that the Metroplis–Hastings method with axillary functions outperforms the standard Metroplis–Hastings method. We plan to test convexization methods on a broad class of MCMC-based methods and develop a detailed methodology for a dynamic adjustment of scaling parameters for auxiliary functions in iterative MCMC processes.

References

[1] L. Bottou, Stochastic learning, in: O. Bousquet and U. von Luxburg (eds.), Advanced Lectures on Machine Learning, pp. 146–148, in: O. Bousquet et al. (eds.), *Machine Learning*, Lecture Notes in Artificial Intelligence LNAI 3176, Springer, Berlin Heidelberg, 2004.

[2] E. J. Candes, J. Romberg, and T. Tao, Robust Uncertainty Principles: Exact Signal Reconstruction from Highly Incomplete Frequency Information, *IEEE Transactions on Information Theory* **52** (2006), 489–509.

[3] A. Carmi, P. Gurfil and D. Kanevsky, Methods for Sparse Signal Recovery Using Kalman Filtering with Embedded Pseudo-Measurement Norms and Quasi-Norms, *IEEE Transactions on Signal Processing* **58**(4) (2010), 2405–2409.

[4] V. Goel and P. Olsen, Acoustic Modeling Using Exponential Families, in: *Proceedings of INTERSPEECH-2009*, pp. 1423–1426, 2009.

[5] D. Kanevsky, *Convexization of Functions*, IBM Research Report, 2011.

[6] D. Kanevsky, A. Carmi, L. Horesh, P. Gurfil, B. Ramabhadran and T. Sainath, Kalman Filtering for Compressed Sensing, in: *The 13th International Conference on Information Fusion*, Edinburgh, UK, July 26–29, 2010.

[7] D. Kanevsky, D. Nahamoo, T. N. Sainath, B. Ramabhadran, and P. A. Olsen, A-Functions: A Generalization of Extended Baum-Welch Transformations to Convex Optimization, in: *Proc. of IEEE International Conference on Acoustics, Speech and Signal Processing (ICASSP)*, May 22–27, 2011, pp. 5164–5167. ISSN: 1520-6149.

Author information

Dimitri Kanevsky, IBM T. J. Watson, Yorktown Heights, NY, USA.
Email: `kanevsky@us.ibm.com`

Avishy Carmi, Department of Mechanical and Aerospace Engineering, Nanyang Technological University, Singapore.
Email: `acarmi@ntu.edu.sg`

Chapter 15

Value Simulation of the Interacting Pair Number for Solution of the Monodisperse Coagulation Equation

Mariya Korotchenko and Aleksandr Burmistrov

Abstract. The monodisperse Smoluchowski equation with linear coefficients is considered. We construct value algorithms and analyze their efficiency for estimating the total monomer concentration, as well as the total monomer and dimer concentrations in ensemble, which is governed by the equation under study. A considerable gain in computational costs is achieved via approximate value simulation of the time between interactions combined with the value simulation of the interacting pair number.

Keywords. Monte Carlo Method, Value Estimator, Parametric Analysis.

Mathematics Subject Classification 2010. 65R20, 65C05.

15.1 Introduction

In this chapter we consider the monodisperse Smoluchowski equation, which describes coagulation processes of particles with positive integer sizes. Let the probability of interaction (or *collision*) for particles with sizes i and j during a time interval Δt and for given coagulation coefficients K_{ij} be equal to $K_{ij}\Delta t$. Under these notations, a concentration of l-sized particles $n_l(t)$ at the instant t in the spatially homogeneous case satisfies the following kinetic equation:

$$\frac{\partial n_l(t)}{\partial t} = \frac{1}{2}\sum_{i+j=l} K_{ij} n_i(t) n_j(t) - \sum_{i \geq 1} K_{il} n_i(t) n_l(t). \tag{15.1}$$

Adding initial data $n_l(0) = n_0(l)$, $l > 0$, to Equation (15.1), we obtain a Cauchy problem for the nonlinear Smoluchowski equation. Further, we will estimate the linear functionals of the function $n_l(t)$, and for this purpose value modifications of statistical simulation will be developed.

For numerical estimation of linear functionals we are going to consider evolution of a multiparticle system [5]. For this purpose we will simulate a homogeneous Markov chain whose transitions are due to elementary pair interactions. Further we will use the following notations:

- N_0 is the initial number of particles in the system, given at instant $t = 0$;
- l_i is the size of the particle with number i;
- $k(l_i, l_j \to l) = N_0^{-1} K_{l_i,l_j} \delta_{l_i+l_j,l}$;
- $N \leq N_0$ is the current number of particles in the system;
- number $\varpi = (i, j)$ is the interacting pair;
- $A(X) = \sum_{i=1}^{N-1} \sum_{j=i+1}^{N} a(N, l_i, l_j)$, where $a(\varpi) \equiv a(N, l_i, l_j) = \sum_{l=1}^{\infty} k(l_i, l_j \to l)$;
- $X = (N, L_N) = (N, l_1, \cdots, l_N)$ describes the phase state of the system;
- $P(X, t)$ is the set of probabilities which determines the state distribution of the system at the instant t.

Under a molecular chaos assumption one can obtain in the limit (see [8] for details)

$$\frac{1}{N_0} \sum_{N=1}^{\infty} \sum_{l_2=1}^{\infty} \cdots \sum_{l_N=1}^{\infty} NP(N, l, l_2 \ldots, l_N, t) \to n_l(t), \text{ when } N_0 \to \infty.$$

This limit allows us to estimate solution to Equation (15.1) with the help of linear integral equation in the way described as follows.

We develop weight modifications on the basis of the technique suggested in [7], which introduces the pair number ϖ to the set of phase coordinates. This approach allowed [7] to derive a special integral equation for the function $F(X, \varpi, t) = a(\varpi)P(X, t)$ in the transformed phase space $\mathbf{Z} \times [0, T]$:

$$F(Z, t) = \int_0^t \int_{\mathbf{Z}} F(Z', t') K(Z', t' \to Z, t) \, \mathrm{d}Z' \, \mathrm{d}t' + F_0(Z)\delta(t).$$

Here we used the following notations: $Z = (X, \varpi)$, $\mathrm{d}Z = \mathrm{d}X \, \mathrm{d}\mu_0(\varpi)$. Integration with respect to the measure μ_0 implies summation over all possible pairs ϖ, and integration over $\mathrm{d}X$ means summation over all values of N and L_N. The latter equation can be used to construct standard weight modifications of the statistical simulation for a multiparticle system due to the multiplicative structure of its kernel $K(Z', t' \to Z, t) = K_1(t' \to t|X') K_2(\varpi|X') K_3(X' \to X|\varpi)$. The distribution density of the time between elementary interactions is exponential: $K_1(t' \to t|X') = A(X') \exp\{-A(X')(t - t')\}$. The probability that a pair of particles $\varpi = (i, j)$ interacts in the N-particle system is $K_2(\varpi|X') = a'(\varpi)/A(X')$. Finally, the function $K_3(X' \to X|\varpi)$ defines the transformation of the system after an interaction of the pair ϖ, which results in the replacement of interacting particles i and j with a single particle of the size $l = l_i + l_j$, so $N = N' - 1$.

Usually the following functionals are of interest: $J_H(T) = \int H(X)P(X,T)\,\mathrm{d}X$. For $\tilde{H}(X,t) = H(X)\exp\{-A(X)t\}$, $H(X) \in L_\infty$, the following equality was derived [7]:

$$J_H(T) = \int_0^T\int_{\mathbf{Z}} \tilde{H}(X, T-t')F(Z,t')\,\mathrm{d}Z\,\mathrm{d}t' = (F, \tilde{H}),$$

which we will make use of later. In this chapter we are interested in the estimation of the monomer concentration, i. e., the functional $J_{H_1}(T)$ with $H_1(X) = \sum_{i=1}^{N}\delta(l_i - 1)/N_0$, as well as the monomer and dimer concentration, i. e., the functional $J_{H_{12}}(T)$ with $H_{12}(X) = \sum_{i=1}^{N}[\delta(l_i - 1) + \delta(l_i - 2)]/N_0$.

Note, that these functionals estimate the solution to the initial Equation (15.1): $J_{H_1}(T) = n_1(T) + \mathcal{O}(N_0^{-1})$, $J_{H_{12}}(T) = n_1(T) + n_2(T) + \mathcal{O}(N_0^{-1})$. The deterministic error of order $\mathcal{O}(N_0^{-1})$ occurs due to the finiteness of N_0 (see [8] for details).

15.2 Value Simulation for Integral Equations

This section presents algorithms of the value simulation for estimation of the functionals $J_{H_1}(T)$ and $J_{H_{12}}(T)$.

Let us define another Markov chain $\{Z_n, t_n\}_{n=0}^{\nu}$; $\nu = \max_n\{n : t_n < T\}$ with a transition density $P^*(Z', t' \to Z, t) = P_1(t' \to t|X')P_2(\varpi|X')K_3(X' \to X|\varpi)$ and a distribution density $P_0(Z)\delta(t)$ of the initial state (Z_0, t_0). Then we can define random weights by the formulas

$$Q_0 = F_0(Z)/P_0(Z), \quad Q_n = Q_{n-1}Q(Z_{n-1}, t_{n-1}; Z_n, t_n),$$

$$Q(Z', t'; Z, t) = Q_t \cdot Q_\varpi = \frac{K_1(t' \to t|X')}{P_1(t' \to t|X')} \cdot \frac{K_2(\varpi|X')}{P_2(\varpi|X')}.$$

In order to estimate the functional $J_H(T)$, the "weight" collision estimator ξ can be used (see [9]):

$$\xi = \sum_{n=0}^{\nu} Q_n \tilde{H}(X_n, T - t_n).$$

Using the results of [6] we can obtain the following theorem.

Theorem 15.1. *If $P_0(Z) \neq 0$ for $F_0(Z) \neq 0$ and $Q(Z', t'; Z, t) < +\infty$ for $Z', Z \in \mathbf{Z}$ and $t', t < T$, then $\mathbf{E}\xi = J_H(T)$. Moreover, if the weights are uniformly bounded and $H \in L_\infty$, then there exists T^* such that $\mathbf{V}\xi < +\infty$ whenever $T < T^*$.*

Our aim is to minimize the variance of the estimator ξ. So we suggest using "value" simulation, i. e., using the value function (see [9] for details) to choose better transition

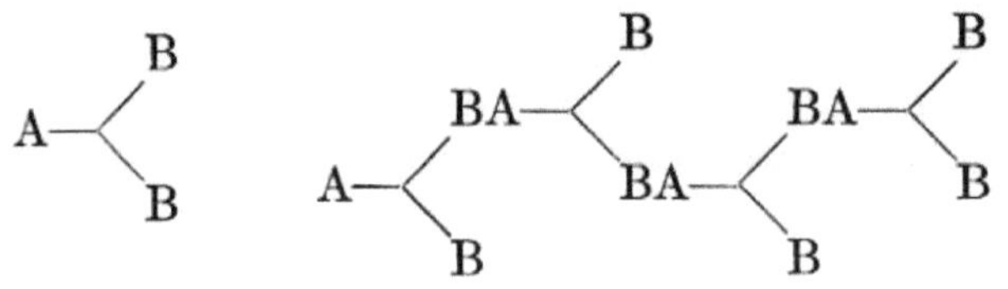

Figure 15.1. Monomer and 4-mer for $f = 3$.

densities P_1 and P_2. Though the value function is usually unknown, we can use the fact that for the problems under consideration, value function is proportional to the number of monomers N_1 (for H_1), or to the sum of monomers and dimers $N_1 + N_2$ (for H_{12}) [4].

Construction of the value algorithm strongly depends not only on a functional for estimation, but also on coagulation coefficients. In this chapter we will illustrate our methodology by the case of linear coefficients $K_{ij} = \mathbf{a} + \mathbf{b}(i + j)/2$. These coefficients can be found, for example, in the classical polymer model $\mathbf{A}—R—\mathbf{B}_{f-1}$ (see [1]). In this model the molecules with $f - 1$ chemically active units of one kind (**B**), and one unit of another kind (**A**), are regarded as monomers. Between **A** and **B** units chemical bonds are formed, which leads to branched molecules if $f \geq 3$ (see Figure 15.1). For this model the coagulation (or polymerization) rate $K_{ij}^{(f)}$ is proportional to $(i + j)(f - 2) + 2$. For the considered case of linear coefficients K_{ij} we have $a(\varpi) = [2\mathbf{a} + \mathbf{b}(l_i + l_j)]/2N_0$, and $A(X) = [\mathbf{a}N/N_0 + \mathbf{b}]\,(N - 1)/2$. Note that this work presents a generalization of the value algorithm suggested previously for constant ($K_{ij} = 1$) and additive ($K_{ij} = [i + j]/2$) coagulation coefficients [2,3]. Moreover, it is possible to implement the parametric analysis (with respect to parameters **a** and **b**) using the standard weight Monte Carlo technique [6] for algorithms proposed below.

The simulation process of the next interaction in the Markov chain includes two successive elementary transitions. First we choose the time interval between interactions, and then we choose two particles for interaction. Let us now describe each of these transitions in detail.

15.2.1 Value Simulation of the Time Interval Between Interactions

For the first elementary transition we suggest using an exponential approximation to the time value function from [2]. This value function was obtained for the case of constant coagulation coefficients ($\mathbf{a} = 1, \mathbf{b} = 0$). Using this approach, we succeeded in constructing efficient value estimators in the case of additive coagulation coefficients ($\mathbf{a} = 0, \mathbf{b} = 1$) [3].

The next elementary transition in our simulation process is the *Value Simulation of the Interacting Pair Number* (*VSIPN*). This stage depends on the type of the estimated functional and is described below for functionals under study.

15.2.2 VSIPN to Estimate the Monomer Concentration J_{H_1}

Let us denote N' to be the total number of particles and N_1' to be the number of monomers before the choice of ϖ. In this case $H(X) = N_1/N_0$.

The VSIPN algorithm suggested below aims at preservation of the monomers in the simulated ensemble. It results in a better estimation of the monomer concentration at the instant T.

First we assign each possible interacting pair of particles to one of three nonoverlapping subsets: $\varpi \in \varpi_1 \cup \varpi_2 \cup \varpi_0$. The choice of the subset depends on the change in the number of monomers takeing place after the collision. The number of monomers may decrease by one (interaction of 'minus-1-pairs'), by two (interaction of 'minus-2-pairs'), or may not change (interaction of 'minus-0-pairs'):

$\boldsymbol{\varpi_1}$ contains 'minus-1-pairs' of the form {*monomer, multimer*};

$\boldsymbol{\varpi_2}$ contains 'minus-2-pairs' of the form {*monomer, monomer*};

$\boldsymbol{\varpi_0}$ contains 'minus-0-pairs' of the form {*multimer, multimer*}.

Note that here by a multimer we mean an l-sized particle where $l \geq 2$. Further, we introduce a representation of the "physical" distribution density $\mathcal{P}_0(i,j) = \alpha[2\mathrm{a} + \mathrm{b}(l_i + l_j)]$ (here $1/\alpha = [N' - 1][\mathrm{a}N' + \mathrm{b}N_0]$) of the interacting pair number in the randomized form

$$1 \equiv \sum_{\varpi} \mathcal{P}_0(i,j) = p_1 \sum_{\varpi_1} f_1(i,j) + p_2 \sum_{\varpi_2} f_2(i,j) + p_0 \sum_{\varpi_0} f_0(i,j). \quad (15.2)$$

Here p_{m} is the probability for choosing the subset ϖ_{m}, and $f_{\mathrm{m}}(i,j)$ is the probability forchoosing the pair (i,j) from ϖ_{m}, $\mathrm{m} = 0, 1, 2$:

$$p_0 = \alpha(N' - N_1' - 1)[\mathrm{a}(N' - N_1') + \mathrm{b}(N_0 - N_1')], \quad (15.3)$$
$$p_1 = \alpha N_1'[2\mathrm{a}(N' - N_1') + \mathrm{b}(N_0 + N' - 2N_1')], \quad p_2 = \alpha N_1'(N_1' - 1)[\mathrm{a} + \mathrm{b}].$$

The monomers are chosen uniformly within the subsets ϖ_1 and ϖ_2. Multimers are chosen within the subsets ϖ_0 and ϖ_1 by their "physical" probabilities $\mathcal{P}_j$, $j = N_1' + 1, \ldots, N'$, having following form:

- for pairs from ϖ_1: $\mathcal{P}_j = \dfrac{2\mathrm{a}+\mathrm{b}(1+l_j)}{2\mathrm{a}(N'-N_1')+\mathrm{b}(N_0+N'-2N_1')}$;
- for pairs from ϖ_0: $\mathcal{P}_j = \dfrac{2\mathrm{a}(N'-N_1'-1)+\mathrm{b}(N_0-N_1')+\mathrm{b}l_j(N'-N_1'-2)}{2(N'-N_1'-1)[\mathrm{a}(N'-N_1')+\mathrm{b}(N_0-N_1')]}$.

In order to "preserve" the monomers, let us carry out the simulation according to (15.2), replacing the probabilities p_{m} from (15.3) by the probabilities q_{m}. Note, that q_{m} is proportional to the number of monomers left in the system:

$$q_1 = p_1 \left(N_1' - 1\right) C_m, \quad q_2 = p_2 \left(N_1' - 2\right) C_m, \quad q_0 = p_0 N_1' C_m,$$
$$1/C_m = \mathbf{E}(N_1) = \alpha N_1'(N' - 2)\left(\mathrm{a}N' - \mathrm{a} + \mathrm{b}N_0 - \mathrm{b}\right).$$

Such modification is taken into account by the multiplicative weight $Q_{\varpi} = p_{\mathrm{m}}/q_{\mathrm{m}}$.

15.2.3 VSIPN to Estimate the Monomer and Dimer Concentration $J_{H_{12}}$

For estimation of functional $J_{H_{12}}$ we will use the technique described in Section 15.2.2 with some necessary modifications. Let us denote N_2' to be the number of dimers before the choice of ϖ and $N_{12}' = N_1' + N_2'$. In this case $H(X) = (N_1 + N_2)/N_0$. We introduce the distribution density proportional to the quantity $N_1(X) + N_2(X)$ in order to introduce VSIPN. Further on we will refer to the l-sized particle with $l \geq 3$ as a multimer.

Taking this into account, let us split the set of all possible interacting pairs into six nonoverlapping subsets: $\varpi_{11} \cup \varpi_{1k} \cup \varpi_{2k} \cup \varpi_{22} \cup \varpi_{12} \cup \varpi_{kk}$. The choice of the subset is related to the change in the total number of monomers and dimers, which results from the interaction (this quantity may decrease by one or two, or not change):

$\boldsymbol{\varpi_{11}}$ contains 'minus-1-pairs' of the form {*monomer, monomer*};

$\boldsymbol{\varpi_{1k}}$ contains 'minus-1-pairs' of the form {*monomer, multimer*};

$\boldsymbol{\varpi_{2k}}$ contains 'minus-1-pairs' of the form {*dimer, multimer*};

$\boldsymbol{\varpi_{22}}$ contains 'minus-2-pairs' of the form {*dimer, dimer*};

$\boldsymbol{\varpi_{12}}$ contains 'minus-2-pairs' of the form {*monomer, dimer*};

$\boldsymbol{\varpi_{kk}}$ contains 'minus-0-pairs' of the form {*multimer, multimer*}.

Let us represent the "physical" distribution density $\mathcal{P}_0$ of the interacting pair number in the form, similar to (15.2)

$$\begin{aligned} 1 \equiv \sum_{\varpi} \mathcal{P}_0(i,j) = p_{11} \sum_{\varpi_{11}} f_{11}(i,j) + p_{1k} \sum_{\varpi_{1k}} f_{1k}(i,j) + p_{2k} \sum_{\varpi_{2k}} f_{2k}(i,j) \\ + p_{12} \sum_{\varpi_{12}} f_{12}(i,j) + p_{22} \sum_{\varpi_{22}} f_{22}(i,j) + p_{kk} \sum_{\varpi_{kk}} f_{kk}(i,j), \end{aligned} \tag{15.4}$$

where p_{mn} is the probability to choose the subset ϖ_{mn}, and $f_{\mathrm{mn}}(i,j)$ is the probability to choose the pair (i,j) from ϖ_{mn}, m, n $\in \{1,2,k\}$. Note, that

$$\begin{aligned} p_{11} &= p_2, \qquad p_{12} = \alpha N_1' N_2' [2\mathrm{a} + 3\mathrm{b}], \qquad p_{22} = \alpha N_2'(N_2' - 1)[\mathrm{a} + 2\mathrm{b}], \\ p_{1k} &= \alpha N_1' [2\mathrm{a}(N' - N_{12}') + \mathrm{b}(N_0 + N' - 2N_{12}' - N_2')], \\ p_{2k} &= \alpha N_2' [2\mathrm{a}(N' - N_{12}') + \mathrm{b}(N_0 + 2N' - 3N_{12}' - N_2')], \\ p_{kk} &= \alpha(N' - N_{12}' - 1)[\mathrm{a}(N' - N_{12}') + \mathrm{b}(N_0 - N_{12}' - N_2')]. \end{aligned} \tag{15.5}$$

Monomers and dimers are chosen uniformly within the subsets $\varpi_{11}, \varpi_{1k}, \varpi_{2k}, \varpi_{22}$ and ϖ_{12}. The multimers are chosen within the subsets ϖ_{1k}, ϖ_{2k} and ϖ_{kk} according to "physical" probabilities $\mathcal{P}_j$, $j = N_{12}' + 1, \ldots, N'$, which have the following form:

- for pairs from ϖ_{1k}: $\mathcal{P}_j = \dfrac{2\mathrm{a} + \mathrm{b}(1 + l_j)}{2\mathrm{a}(N' - N_{12}') + \mathrm{b}(N_0 + N' - 2N_{12}' - N_2')}$;

- for pairs from ϖ_{2k}: $\mathcal{P}_j = \dfrac{2\mathrm{a}+\mathrm{b}(2+l_j)}{2\mathrm{a}(N'-N'_{12})+\mathrm{b}(N_0+2N'-3N'_{12}-N'_2)}$;
- for pairs from ϖ_{kk}: $\mathcal{P}_j = \dfrac{2\mathrm{a}(N'-N'_{12}-1)+\mathrm{b}[(N_0-N'_{12}-N'_2)+l_j(N'-N'_{12}-2)]}{2(N'-N'_{12}-1)[\mathrm{a}(N'-N'_{12})+\mathrm{b}(N_0-N'_{12}-N'_2)]}$.

In order to "preserve" the monomers and dimers, we will choose the interacting pair according to (15.4) with the probabilities (15.5) replaced by q_{mn}, proportional to the sum of the monomers and dimers left in the system:

$$q_{11} = (N'_{12}-1)p_{11}C_{md};\quad q_{1k} = (N'_{12}-1)p_{1k}C_{md};\quad q_{2k} = (N'_{12}-1)p_{2k}C_{md};$$
$$q_{12} = (N'_{12}-2)p_{12}C_{md};\quad q_{22} = (N'_{12}-2)p_{22}C_{md};\quad q_{kk} = (N'_{12}-0)p_{kk}C_{md},$$

where $C_{md}^{-1} = \mathbf{E}(N_{12}) = \alpha N'_{12}(N'-2)\,(\mathrm{a}N'-\mathrm{a}+\mathrm{b}N_0-2\mathrm{b}) + \alpha N'_1[\mathrm{a}(N'_1-1)+\mathrm{b}(N'+N'_1-3)]$. This modification is taken into consideration when the weight is calculated: $Q_\varpi = p_{\mathrm{mn}}/q_{\mathrm{mn}}$.

15.3 Results of the Numerical Experiments

In this section the simulation results according to the suggested algorithms are presented and compared with the analytic solution of the test problem. As a test problem for implementation of the algorithms described above, we take the problem (15.1) with $K_{ij} = \mathrm{a}+\mathrm{b}(i+j)/2$, $n_0(l) = \delta_{l,1}$. This problem has an exact solution in the form (see [10]):

$$n_l(\mu(t)) = \mu(1-\mu)^{l-1}\left(\frac{\mathrm{a}\mu+\mathrm{b}}{\mathrm{a}+\mathrm{b}}\right)^{1+l\mathrm{b}/\mathrm{a}}, \qquad \mu(t) = \frac{\mathrm{b}}{(\mathrm{a}+\mathrm{b})\exp\{\mathrm{b}t/2\}-\mathrm{a}}.$$

We used the following notations in the tables:

- $\bar{\sigma}$ is the mean square error (square root of the estimate variance);
- PE (%) is the percent error;
- $t^{(c)}$ is the calculation time;
- M is the number of simulated trajectories;
- $\mathrm{S}_d = \bar{\sigma}_d^2 t_d^{(c)}$ and $\mathrm{S}_v = \bar{\sigma}_v^2 t_v^{(c)}$ are the computational costs for the direct and value simulation, respectively.

Numerical results show that there is a considerable reduction in computational cost when we use value simulation for both elementary transitions in a Markov chain. This effect takes place for various parameters **a** and **b**. Moreover, gain in computational cost is even more noticeable for large values of T. You can find some of the results in Tables 15.1–15.2. Furthermore, with the help of the developed algorithms we succeeded in the estimation of small values for which direct simulation fails (see Table 15.3).

Table 15.1. Estimation of $J_{H_1}(T)$ for $\mathbf{a} = 1$, $\mathbf{b} = 2$ ($T = 1; 10$).

Simulation	$\tilde{J}_{H_1}(T)$	$\bar{\sigma}$	*PE (%)*	t_c	S_d/S_v
	$n_1(1) = 1.2263 \cdot 10^{-1}$ ($T = 1$; $M = 10^4$; $N_0 = 10^3$)				
direct	$1.2266 \cdot 10^{-1}$	$1.1 \cdot 10^{-4}$	0.02	37.7	––
value	$1.2271 \cdot 10^{-1}$	$4.7 \cdot 10^{-5}$	0.07	34.9	**6.09**
	$n_1(10) = 8.9684 \cdot 10^{-6}$ ($T = 10$; $M = 10^5$; $N_0 = 2 \cdot 10^2$)				
direct	$9.4500 \cdot 10^{-6}$	$6.9 \cdot 10^{-7}$	5.37	68.9	––
value	$8.6176 \cdot 10^{-6}$	$5.4 \cdot 10^{-8}$	3.91	74.9	**152**

Table 15.2. Estimation of $J_{H_{12}}(T)$ for $\mathbf{a} = 1$, $\mathbf{b} = 2$ ($T = 1; 10$).

Simulation	$\tilde{J}_{H_{12}}(T)$	$\bar{\sigma}$	*PE (%)*	t_c	S_d/S_v
	$n_1(1) + n_2(1) = 1.7364 \cdot 10^{-1}$ ($T = 1$; $M = 10^4$; $N_0 = 10^3$)				
direct	$1.7379 \cdot 10^{-1}$	$1.4 \cdot 10^{-4}$	0.08	25.9	––
value	$1.7388 \cdot 10^{-1}$	$6.8 \cdot 10^{-5}$	0.13	28.9	**3.58**
	$n_1(10) + n_2(10) = 1.2954 \cdot 10^{-5}$ ($T = 10$; $M = 10^5$; $N_0 = 2 \cdot 10^2$)				
direct	$1.3600 \cdot 10^{-5}$	$8.3 \cdot 10^{-7}$	5.37	64.1	––
value	$1.2594 \cdot 10^{-5}$	$7.2 \cdot 10^{-8}$	2.78	72.5	**117**

Table 15.3. Estimation of $J_{H_1}(T)$ and $J_{H_{12}}(T)$ for $\mathbf{a} = 2$, $\mathbf{b} = 10$.

Simulation	$\tilde{J}_H(T)$	$\bar{\sigma}$	*PE (%)*
	$n_1(10) = 5.3828 \cdot 10^{-23}$ ($T = 10$; $M = 10^6$; $N_0 = 10^3$)		
value	$5.3047 \cdot 10^{-23}$	$1.9 \cdot 10^{-25}$	1.45
direct	**0.0**	––	**100**
	$n_1(10) + n_2(10) = 7.5460 \cdot 10^{-23}$ ($T = 10$; $M = 10^6$; $N_0 = 10^3$)		
value	$7.4278 \cdot 10^{-23}$	$2.1 \cdot 10^{-25}$	1.57
direct	**0.0**	––	**100**

15.4 Conclusion

The proposed method not only allows reduction of the computational cost of estimating the functionals, but also makes it possible to implement the parametric analysis (with respect to parameters **a** and **b**) using the standard weight Monte Carlo technique.

Acknowledgments. This work was partly supported by RFBR (grant 11-01-00252) and SB RAS (Integration Grant No. 47). The authors acknowledge the kind hospitality of the Institute of Information and Communication Technologies, BAS, and the Eighth IMACS Seminar MCM-2011 organizers.

References

[1] P. J. Flory, *Principles of polymer chemistry.* Cornell University Press, Ithaca, NY, 1953.

[2] M. A. Korotchenko, Value modifications of statistical algorithms for solution of the Smoluchowski equation, *Vych. Tekhnol.* **13**(sp.iss. 4) (2008), 68–74 (in Russian).

[3] M. A. Korotchenko, *Value Monte Carlo algorithms for estimating the solution to the coagulation equation.* In: L. Plaskota and H. Wozniakowski (eds.), *Monte Carlo and Quasi-Monte Carlo Methods 2010*, pp. 511-522, Springer Proceedings in Mathematics & Statistics 23, Springer-Verlag, Heidelberg, 2012.

[4] M. A. Korotchenko, G. A. Mikhailov, and S. V. Rogasinsky, Modifications of weighted Monte Carlo algorithms for nonlinear kinetic equations, *Comp. Math. Math. Phys.* **47**(12) (2007), 2023–2033.

[5] A. A. Lushnikov, Some new aspects of coagulation theory, *Izv. Akad. Nauk SSSR, Ser. Fiz. Atmosfer. i Okeana* **14**(10) (1978), 738–743 (in Russian).

[6] G. A. Mikhailov, *Parametric Estimates by the Monte Carlo Method.* VSP, Utrecht, 1999.

[7] G. A. Mikhailov and S. V. Rogasinsky, Weighted Monte Carlo methods for approximate solution of the nonlinear Boltzmann equation, *Sib. Math. J.* **43**(3) (2002), 496–503.

[8] G. A. Mikhailov, S. V. Rogasinsky, and N. M. Ureva, Weighted Monte Carlo methods for an approximate solution of the nonlinear coagulation equation, *Comp. Math. Math. Phys.* **46**(4) (2006), 680–690.

[9] G. A. Mikhailov and A. V. Voitishek, *Numerical Statistical Modelling (Monte Carlo Method).* Akademia, Moscow, 2006 (in Russian).

[10] J. L. Spouge, Solutions and critical times for the monodisperse coagulation equation when $a(i,j) = A + B(i+j) + Cij$, *J. Phys. A: Math. Gen.* **16** (1983), 767–773.

Author information

Mariya Korotchenko, Institute of Computational Mathematics and Mathematical Geophysics, Siberian Branch of the Russian Academy of Sciences, Novosibirsk, Russia.
Email: kmaria@osmf.sscc.ru

Aleksandr Burmistrov, Institute of Computational Mathematics and Mathematical Geophysics, Siberian Branch of the Russian Academy of Sciences, Novosibirsk, Russia.
Email: burm@osmf.sscc.ru

Monte Carlo Methods and Applications, 145–153

Chapter 16

Parallelization of Algorithms for Solving a Three-Dimensional Sudoku Puzzle

Mikhail Mayorov and Paula A. Whitlock

Abstract. The well-known logic puzzle Sudoku can be generalized from two to three dimensions by designing a puzzle that is played on the six faces of a cube. Three-dimensional Sudoku puzzles with two variations of the constraints for the edges of the cube have been studied using the method of simulated annealing. The computer code to solve the puzzle is parallelized using MPI.

Keywords. Sudoku, Simulated Annealing, Markov Chains, MPI.

Mathematics Subject Classification 2010. 65C05, 68M14, 68W15.

16.1 Introduction

Previously, Lambert and Whitlock [1] introduced the concept of solving three dimensional Sudoku puzzles using a Monte Carlo method. Their work was based on Lewis [2], who solved the standard two-dimensional puzzle by simulated annealing [3] and found interesting behavior of the algorithm as the number of fixed cells in the puzzle changed. Lambert and Whitlock also observed a change in the behavior of the simulated annealing process on a puzzle covering three faces of a cube. However, the code written to solve puzzles on all six faces of a cube proved to be a computationally challenging problem using a serial computer program. The research described here extends the investigation of three-dimensional sudoku puzzles by parallelizing a computer code using MPI.

The two-dimensional Sudoku puzzle is well known and is played in many environments, such as newspapers, on the internet, as a phone app, etc. It is a logic puzzle consisting of an $n^2 \times n^2$ grid of cells partitioned into n^2 $n \times n$ blocks. When successfully completed, every row, column, and block in the puzzle must contain the numbers $1 - n^2$ just once. Games with $n = 3$ and $n = 4$ are commonly available. The general problem of solving Sudoku puzzles is known to be NP-complete [4], and many methods have been developed to obtain optimal solutions. However, the Sudoku puzzles people usually play are called logic-solvable because a logical chain of reasoning will lead to a solution. Such puzzles come with some cells having assigned values, called fixed cells, and the empty cells are to be filled in by the player. But there are also puzzles whose solution can only be found by guessing a random solution or by applying brute force iteration through all possible combinations. A good example of such a puzzle is an initially empty puzzle, with no fixed cells at all.

Solving a Sudoku puzzle on the six faces of a cube brings additional complexity. Not only should each puzzle be solved on each face of a cube, but all faces must be consistent with each other: border cells on the edges of adjacent faces must either strictly match or strictly not match. In other words, solving a puzzle on any one face cannot be done independently from the other faces, because each change on a border cell of a face leads to a several changes on one or more adjacent faces. In this research, several algorithms for solving the six-face puzzle were tried. All of them involved the method of simulated annealing and were distributed across a network using MPI.

16.2 The Simulated Annealing Method

To solve an optimization problem by simulated annealing, a cost function $U(x)$ is used to measure how close a proposed solution $s(x)$ is to an optimal solution. The probability distribution of the proposed solutions is given by

$$F(X) \propto e^{-t \cdot U(X)}, \tag{16.1}$$

where t is labelled an "inverse temperature." The cost function has either a maximum or a minimum when an optimal solution is located. The search algorithm begins at a "high temperature", which allows large excursions into the initial search space. As the cost function value changes, the value of t changes according to a cooling schedule.

Both the previous investigation [1] and the work described here uses the basic approach introduced by Lewis [2]. As the simulating annealing algorithm proceeds, a "neighborhood operator" is applied. The operator chooses two different nonfixed cells in the same block and swaps them. A sampling bias towards blocks with fewer nonfixed values is prevented by the following steps:

1. Choose randomly i and j, such that
   (a) $1 \le i, j \le n^2$, and
   (b) cell $[i][j]$ is nonfixed.
2. Choose randomly k and l, such that
   (a) the cell $[k][l]$ is in the same block as cell $[i][j]$, (2.2)
   (b) cell $[k][l]$ is nonfixed,
   (c) cell $[k][l] \neq$ cell $[i][j]$.
3. Swap cell $[i][j]$ with cell $[k][l]$.

To compute the cost function $U(x)$ each row is evaluated and a sum of the values $1-n^2$ which are *not* present in the row is calculated. The same calculation is done for each column, and the $U(x)$ is the total sum of these values. An optimal solution will have a cost of zero.

A single application of the neighborhood operator, Algorithm (2.2), means that at most two rows of cells and two columns of cells are affected by a swap. After a single move in the block, only the contributions to the cost function that have been altered are recalculated. This efficient evaluation of the cost function is called the delta-evaluation [5] and leads to a considerable speed-up of the algorithm.

Starting with an initial candidate solution, $s(x)$, an exploration of the search space is conducted by iteratively applying the neighborhood operator, Algorithm (2.2). A new possible solution $s(x')$ is accepted i) if the cost function $U(x')$ associated with $s(x')$ is smaller than $U(x)$ with respect to $s(x)$, or ii) with probability $e^{(-\delta(U(x)-U(x'))\cdot t)}$ where $\delta(U(x) - U(x'))$ is the proposed change in the cost function. Proposed changes in $s(x')$ that satisfy neither i) nor ii) above are rejected.

The way in which t is altered during the run is of great importance to the success and timing of the simulated annealing algorithm. It is generally believed that the initial temperature, t_0, should allow the majority of moves to be accepted [6]. Subsequently, it should be slowly reduced according to a cooling schedule so that the algorithm becomes increasingly greedy. If the algorithm is successful, eventually the search should converge at or near a global optimum. Van Laarhoven and Aarts [6] advise that t_0 should allow approximately 80 % of the proposed moves to be accepted. This behavior was confirmed by Lambert and Whitlock [1]; see their Table 3.

In the simulation described here, the temperature was reduced using a simple geometric cooling schedule. The current temperature t_i is modified to a new temperature t_{i+1} by $t_{i+1} = \alpha \times t_i$, where α is the cooling rate, $0 < \alpha < 1$.

The algorithm can be summarized as follows. A sequence of Markov chains is formed, where each Markov chain is generated at a fixed value of t. The Markov chain needs to be long enough for sufficient exchanges of cells to occur for the system to equilibrate. As t is altered, a new Markov chain of cell exchanges is performed. When the cost function becomes 0, a solution has been found and the simulation terminates. If the temperature falls below $t < 10^{-6}$ and no solution is found, the temperature is reset to t_0 and the simulation begins again. This behavior is termed a reheat.

16.3 Successful Algorithms for Solving the Three-Dimensional Puzzle Using MPI

While the six-face puzzle can be solved on a single computer using the simulated annealing method, the calculation can be very slow. It was decided to distribute the simulation over a small number of computers on a network using MPI. Several different approaches to distributing the simulation were tried, the two fastest and most successful are described here.

In the computer codes written to solve the Sudoku puzzle, each cell on a face is identified by three numbers. The first number denotes the face number, the second and third numbers, respectively, denote the row and the column within the face. All numeration starts with 0. Thus, for example, cell [3][5][2] represents the cell which is located on the third face in the fifth row and second column. This representation is easily translated into a three-dimensional array. Since the first dimension identifies the face, its size is always 6. The size of the second and third dimensions are the same, n^2, and vary according to the size of the puzzle to be solved. The memory used for the storage is $6 * (n^4) * \mathit{sizeof}(\mathit{int})$ bytes.

After an initial representation of a puzzle is read into the program, an initial candidate solution is created by assigning a value to each nonfixed cell in the array. This is done randomly but consistently, so that on each face, within each block, every cell contains the values 1 to n^2 exactly once.

16.3.1 An Embarrassingly Parallel Algorithm

In this algorithm each independent MPI process tries to solve all six faces of the whole puzzle. This strategy includes the following steps:

1. All border values on all six faces are initially randomly populated in a logically consistent way.
2. Internal cells on each face are filled randomly.
3. Each process starts to solve the puzzle on face #0 using the neighborhood operator (2.2). If after a fixed number of reheats the process cannot find the solution, it randomly picks two border cells and exchanges the values. The change is propagated to all faces so that the borders are consistent.
4. After the border exchange, the simulation continues to solve the puzzle by exchanging values of internal cells. The steps are repeated until the puzzle is solved on face #0.
5. Once the puzzle on the face #0 is solved, the process marks all the face's border cells as fixed. The cells on the edges of adjacent faces become fixed as well if the edges must match.
6. The simulation begins to solve the puzzle on an adjacent face. The same logic is followed as described in step 3 above with one exception. Exchanges of border cells can only occur on nonfixed edges.
7. Steps 3–6 are repeated for every face until the whole puzzle is solved.

The simulation is ended when a process successfully solves the puzzle.

Experiments with initially empty puzzles have shown that fixing the border cells initially with randomly assigned values and not allowing exchanges, step 3 above, leads to puzzles that may not have a solution. When performing these simulations, an interesting phenomenon was observed. As each face is solved and the next face is processed, there is less and less freedom to exchange values on the border cells. In the final stage of the algorithm, when five faces have already been solved, the border cells are all fixed. That is, no change on any border cell of the last face is allowed. At this point, in every experiment performed, the solution always exists and the last face is solved very quickly.

The codes to solve the Sudoku puzzle where run on a network with 86 nodes. Each node has two Intel 2.80 GHz Woodcrest dual-core processors connected by a one Gigabit ethernet. Empty puzzles where each face has dimensions 9×9, a cooling schedule where $\alpha = 0.90$, and the constraint that edge cells must match were solved. On average, it was found that it took 0.444 seconds with a standard deviation of 0.10 when the calculation was distributed among 10 processors.

16.3.2 Distributed Simulated Annealing Using a Master/Worker Organization

While the algorithm described above has communication only at the beginning and at the end of the calculation, the next algorithm uses a master/worker organization with more communication.

1. The master process, process 0,
   (a) assigns values to nonfixed edge cells according to the edge constraint,
   (b) assigns random values to empty internal cells,
   (c) distributes face #0 to all worker processes.
2. The first process to solve face #0 returns the result to the master process. The master process then
   (a) fixes the border cells on adjacent faces,
   (b) distributes the remaining faces to the worker process.
3. Worker processes return faces as they are solved.

Process 0 keeps track of how many faces are solved, and as soon as all faces are solved, the whole puzzle is solved and the program terminates. Even though there is extensive communication, this algorithm is faster than the previous one. For example, starting with an empty puzzle and distributing the calculation among 10 processors, the master/worker code solved the puzzle in 0.114 ± 0.02 seconds on the average. Thus there is a significant decrease in the running time of the calculation over the first algorithm.

16.4 Results

The distributed three-dimensional Sudoku code is not scaleable, that is, there is no advantage to using a larger numbers of CPUs. The limiting factor here is the time

Table 16.1. The percentage of empty cells that correspond to different behaviors of the simulated annealing code.

Constraint	*Percent of empty cells for unique puzzles*	*Degenerate solutions*	*No solution*
Edge cells must match	0–44%	45%–100%	80%–90%
Edge cells must not match	0–29%	30%–100%	35%–95%

it takes to solve an individual face of the puzzle. If solving a face necessitates long sequences of Markov chains and multiple reheats, the time is considerably increased. Several configurations of nodes and cores were tried. It was found that the consistently best timings were achieved by using three nodes on the network with two cores per node. Using more cores on a node increased the time, probably due to cache issues on the Woodcrest dual-core processors.

Another interesting aspect of solving the three-dimensional Sudoku puzzle was the differing behavior with respect to the edge constraints. Repeating the experiment of starting with empty puzzles many times, it was found that the constraint that the edge cells match was solved faster than the case where the edge cells did not match. This behavior results from how the implementation of the simulated annealing algorithm proceeds. When the edges must match, the edge cells on a neighboring face are fixed when a face puzzle is solved. However, in the alternative case, solving a face only specifies the value that cannot be assigned to an adjacent face edge cell. Thus the exchange of border cells becomes necessary, and more possible cells participate in the solution process on the neighboring faces.

Another question is how many fixed cells are needed to produce a unique puzzle, that is, a puzzle that has only one solution when completed. Beginning with a solved six-face puzzle, a percentage of the cells were randomly emptied, i. e., 5 %, 10 %, … , and the puzzle was resolved by the code with a different sequence of pseudorandom numbers. If the new solution was identical to the original puzzle after several repetitions of the process, the solution was deemed unique. The results of these experiments are listed in Table 16.1. For the case that the edges must match and faces are of the size 9×9, if less than 243 out of 486 total cells are empty, the puzzle has a unique solution. For the puzzle with edge cells that must not match, the solution will be unique if less than 145 cells are empty.

The percentage of fixed cells needed for a unique puzzle can be decreased by systemically choosing to remove the values from internal cells rather than from edge cells.

As the number of fixed cells decreases, the time to complete the puzzle increases. Similar behavior was observed by both Lewis [2] and Lambert and Whitlock [1]. In the three-face puzzle case, a maximum time to solve the puzzle was observed for faces of size 16×16 with between 50 % and 70 % of the cells empty. In addition, in some cases the puzzle could not be solved at all in the time allowed. Similar behavior is present in the six-face puzzle. However, the likelihood of the puzzle not being able to

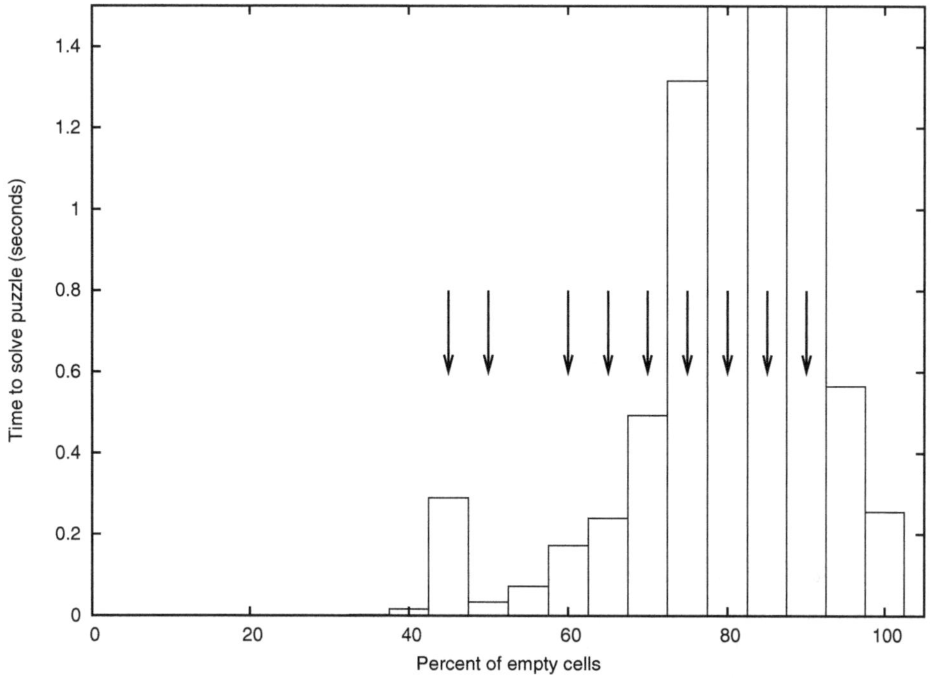

Figure 16.1. Average time to solve a Sudoku puzzle on the faces of a cube when $n = 3$. The arrows indicate puzzles where a solution may not be found in 30 seconds of execution time.

be solved in a fixed amount of time occurs even for the case of faces of size 9×9. If the edge cells must match, the time to solve the puzzle begins to increase dramatically when more than 70 % of the cells are randomly emptied. It was observed that after many repetitions, puzzles with more than 80 % empty cells were not solved in 30 s. If the edge cells must not match, some puzzles were not solved when 35 % of the cells were randomly emptied. By 50 % of the cells empty, no solution was found in 30 s in all the experiments tried. The lack of a solution usually occurs when a large number of reheats occurred, which implies a larger search space and longer times to solve the puzzle. The increased frequency of unsolved puzzles in this case compared with the three-face puzzle [1] is related to the change in the algorithm. Previously, the neighborhood operator was applied simultaneously to all three faces. Here, face #0 is solved first and can create a circumstance where the remaining faces with fixed cells cannot be solved. This will not occur with a completely empty initial puzzle – no fixed cells, or with puzzles that have just one unique solution.

16.5 Conclusions

The simulated annealing codes can be used to solve puzzles and to explore some properties of the three-dimensional Sudoku puzzles such as how many fixed cells need to be present for a puzzle to have a unique solution. However, the issue of whether or not the puzzle can be solved by applying a chain of reasoning is not answered by this algorithm. Several additional distributed codes were written using a backtracking algorithm whose purpose was to generate logically solvable puzzles [7]. Similarly to the simulated annealing algorithm, it was found that the constraint that edge cells do not match results in a puzzle much more difficult to solve.

For the purpose of developing a playable game, several necessary conditions can be specified. The puzzle must have a unique solution. To solve the puzzle, the player should be able to use a chain of reasoning to obtain the unique solution. Probably, the constraint that edge cells must match would lead to more easily solved puzzles. Finally, the player should be able to geometrically envision the game.

Acknowledgments. This research was supported in part under National Science Foundation Grants CNS-0958379 and CNS-0855217 and by the City University of New York High Performance Computing Center. One of us, P. A.W., thanks Professor Marvin Bishop, Manhattan College, for many useful conversations.

References

[1] T. Lambert and P. A. Whitlock, Generalizing Sudoku to Three Dimensions. *Monte Carlo Methods and Applications* **16** (2010), 251–263.

[2] R. Lewis, Metaheuristics can solve Sudoku puzzles. *Journal of Heuristics* **13** (2006), 387–401.

[3] S. Kirkpatrick, C. D. Gelatt, Jr., and M. P. Vecchi, Optimization by Simulated Annealing. *Science* **220** (1983), 671–680.

[4] R. Santos-Garcia and M. Palomino, Solving Sudoku Puzzles with Rewriting Rules, *Electronic Notes in Theoretical Computer Science* **176** (2007), 79–83.

[5] P. Ross, D. Corne, and H.-L. Fang, Improving Evolutionary Timetabling with Delta Evaluation and Directed Mutation. in: . Y. Davidor, H. Schwefel and M. Reinhard (eds.), Parallel Problem Solving From Nature III (PPSN), pp. 556–565, Lecture Notes in Computer Science 866, Springer-Verlag. Berlin Heidelberg, 1994 .

[6] P. van Laarhoven and E. Aarts, *Simulated Annealing: Theory and Applications*, D. Reidel Publishing Company, The Netherlands, 1987.

[7] M. Mayorov, Parallel Implementation of Three-dimensional Sudoku Puzzles using Simulated Annealing and a Backtracking Algorithms, master's thesis. Brooklyn College CUNY, 2011.

Author information

Mikhail Mayorov, Department of Computer and Information Sciences, Brooklyn College, Brooklyn, NY, USA.
Email: mikhailmayorov83@gmail.com

Paula A. Whitlock, Department of Computer and Information Sciences, Brooklyn College, Brooklyn, NY, USA.
Email: whitlock@brooklyn.cuny.edu

Chapter 17

The Efficiency Study of Splitting and Branching in the Monte Carlo Method

Ilya N. Medvedev

Abstract. The Monte Carlo solution of second-kind integral equations is usually based on the von Neumann–Ulam scheme, since these equations are related to homogeneous Markov chains terminating with probability 1. In this chapter we study the computational cost of the use of the standard "weight" collision estimator and some of its modifications with splitting and branching for estimating the solution of the second-kind integral equation. In general the use of trajectory splitting diminishes the variance of the estimator and increases the computation time. In this chapter we present the optimal value for splitting of the trajectory at a certain instant time in a Markov chain. However, it turned out that the use of trajectory splitting once at the first time in a Markov chain results in greater computational cost as compared with the algorithm without splitting.

The basic characteristic of the weight collision estimator is a random weight that is multiplied by the ratio of the corresponding kernel to the transition density after every transition in the simulated Markov chain. If the weight values do not exceed 1, then the mean square (and, hence, the variance) of the standard collision estimate for the functional is finite. Otherwise we can branch the simulated trajectory to a random number of branches with the expectation that is equal to the weight. We show that the use of such branching provides the greater computational cost as compared with the algorithm without branching.

Keywords. Integral Equation, "Weight" Collision Estimator, Trajectory Splitting and Branching, Computational Efficiency (Cost).

Mathematics Subject Classification 2010. 45B05, 65C05, 65K10.

17.1 Introduction

Let us consider the integral equation

$$\varphi(x) = \int_X k(x, x')\varphi(x')dx' + h(x), \tag{17.1}$$

or $\varphi = K\varphi + h$, where X is a bounded domain in R^n, $h \in L_\infty(X)$, $K \in [L_\infty \to L_\infty]$. Here we assume that the kernel $k(x, x')$ can contain generalized multipliers of the delta function type. In this case the simulated transition distribution density $p(x, x')$ for the Markov chain should contain the same multipliers. Also we consider that the functions $k(x, x')$, $h(x)$ and $\varphi(x)$ are nonnegative and $||K^{n_0}|| < 1$ for some $n_0 \geq 1$.

Using standard technique based on majorization, it is easy to verify that all presented results are valid for estimating functionals (f, φ) where $f \in L_1(X)$ [2,3].

Let us consider a terminating homogeneous Markov chain $x_0, x_1, \ldots, x_N$ with transition distribution density $p(x, x')$ and trajectory termination probability $p(x) = 1 - \int_X p(x, x')dx' \geq 0$ at the transition $x \to x'$. Here and further N is the number of the state at which the trajectory terminates (i. e., the termination moment) [2,3]. The Markov chain is connected with Equation (17.1) by means of the weight multiplier $q(x, x')$ which has the form

$$q(x, x') = \frac{k(x, x')}{p(x, x')}$$

for the transition $x \to x'$. Also we suppose that general unbiasedness condition

$$p(x, x') \neq 0 \quad \text{when} \quad k(x, x') \neq 0$$

is valid [2,3].

We consider the problem of estimating the value $\varphi(x)$ where $\varphi = \sum_{n=0}^{\infty} K^n h$ is the Neumann series for Equation (17.1). For this problem it is convenient to use the weighted "collision estimator"

$$\xi_{x_0} = h(x_0) + \sum_{n=1}^{N} Q_n h(x_n) = \sum_{n=0}^{\infty} \Delta_n Q_n h(x_n),$$

where $Q_0 \equiv 1$, $Q_n = Q_{n-1} q(x_{n-1}, x_n)$ are the weights and Δ_n is the indicator of the nonbreak of the trajectory up to the state x_n. It means that $\Delta_n = 1$ when $n \leq N$ and $\Delta_n = 0$ when $n > N$; $\Delta_0 \equiv 1$. Under the conditions mentioned above we obtain $\mathrm{E}\xi_{x_0} = \varphi(x_0)$ [2,3]. From the definition of ξ_x it follows that

$$\xi_{x_0} = h(x_0) + \Delta_1 q(x_0, x_1) \xi_{x_1}, \tag{17.2}$$

where δ_x is the indicator of non-break of the trajectory at the transition $x \to x'$ and $\delta_{x_0} = \Delta_1$. It is clear that the recurrence $\xi_x = h(x) + \delta_x q(x, x') \xi_{x'}$ uniquely defines the weighted "collision estimator".

If $q(x, x') \leq 1$, then the variance $\mathrm{D}\xi_{x_0}$ is certainly finite [2,3]. Otherwise the value $\mathrm{D}\xi_{x_0}$ can be infinite, and it is impossible to correctly estimate the simulation error. To overcome such a complication we propose to use the appropriate branching of the simulated trajectory.

17.2 Randomized Branching

By the random variable $\nu(x, x')$ with positive integer values we denote the number of branches. We suppose that $\mathrm{E}\nu(x, x') < \infty$.

Let us define the random variable ζ_x by the recurrence

$$\zeta_x = h(x) + \delta_x \frac{q(x,x')}{\mathrm{E}\nu(x,x')} \sum_{i=1}^{\nu(x,x')} \zeta_{x'}^{(i)}, \tag{17.3}$$

where $\{\zeta_{x'}^{(i)}\}$ are the independent realizations of the $\zeta_{x'}$. It is easy to check [4] that under the assumptions mentioned above we have $\mathrm{E}\zeta_{x_0} = \varphi(x_0)$ and the variable $\mathrm{E}\zeta_{x_0}^2$ is defined by the Neumann series for the equation

$$\mathrm{E}\zeta_x^2 = \int_X k(x,x') \frac{q(x,x')}{\mathrm{E}\nu(x,x')} \mathrm{E}\zeta_{x'}^2 dx' + H(x), \tag{17.4}$$

where

$$H(x) = h(x)\{2\varphi(x) - h(x)\} + \int_X k(x,x')\psi(x,x')\varphi^2(x')dx', \tag{17.5}$$

and

$$\psi(x,x') = \frac{q(x,x')\mathrm{E}\{\nu(x,x')(\nu(x,x')-1)\}}{\mathrm{E}(\nu(x,x'))^2}.$$

For the collision estimator under direct simulation ($q(x,x') \equiv 1$) without branching $\xi_x = h(x) + \delta_x \xi_{x'}$ we have the following equation [3]:

$$\mathrm{E}\xi_x^2 = \int_X k(x,x')\mathrm{E}\xi_{x'}^2 dx' + h(x)[2\varphi^*(x) - h(x)]. \tag{17.6}$$

In [4] we studied the possibility of branching the simulated trajectory only if the next weight $q(x,x')$ is < 1. Also we supposed that $\mathrm{E}\nu(x,x') = q(x,x')$. In this case it is easy to check that the Neumann series for $\mathrm{E}\zeta_x^2$ has an additional positive summand $\sum_{n=1}^{\infty}[K^n H](x)$ as compared with $\mathrm{E}\xi_x^2$, and the inequality $\mathrm{D}\zeta_x^2 \geq \mathrm{D}\xi_x^2$ is true.

By the variable S we define the computation cost as

$$S = T\mathrm{D}\xi,$$

the product of the average time T needed for one realization (trajectory) of ξ and the variance estimator $\mathrm{D}\xi$. Let us compare the computation cost for simulation with and without branching.

Let us suppose that the average simulation time t_p ($p(x,x')$) approximately equals the average simulation time t_k ($k(x,x')$) for the one transition $x \to x'$ and $T_{(\cdot)} = t_{(\cdot)}\mathrm{E}N_{(\cdot)}$ where $\mathrm{E}N_{(\cdot)}$ is the average number of the state at which the trajectory terminates (under the use of the simulation density $(\cdot)$).

Let us consider the following equation:

$$n(x) = \int_X k(x,x')n(x')\,dx' + I_{\{x \in X\}}.$$

The solution of the presented equation is $n(x) = \mathrm{E}N_{(k)}(x)$. This is why we can use the estimator $\tilde{\xi}_x = I_{\{x \in X\}} + \delta_x \tilde{\xi}_{x'}$ for estimating the average number of the state at which the trajectory terminates under the use of the simulation density $k(\cdot,\cdot)$ for direct simulation without branching ($\mathrm{E}N_{(k)}(x)$). To estimate the average number of the state at which the trajectory terminates under the use of the simulation density $p(\cdot,\cdot)$ for "weight" simulation with branching we can use the estimator

$$\tilde{\zeta}_x = I_{\{x\in X\}} + \delta_x \sum_{i=1}^{\nu(x,x')} \tilde{\zeta}_{x'}^{(i)}.$$

and the following equality is true:

$$\mathrm{E}\tilde{\zeta}_x = \mathrm{E}N_{(p)}(x) = n(x) = \mathrm{E}N_{(k)}(x) = \mathrm{E}\tilde{\xi}_x.$$

Generalizing the above information , we derive the inequality

$$S_b(x) = \mathrm{D}\zeta_x t_b \mathrm{E}N_{(p)}(x) \geq S_d(x) = \mathrm{D}\xi_x t_d \mathrm{E}N_{(k)}(x),$$

and the fact that the use of presented randomized branching (ζ_x) always results in a greater computation cost compared to direct simulation without branching (ξ_x).

In spite of great computation cost, randomized branching can be applied to problems where direct simulation cannot be applied. For example, let us consider some substance that emits particles in the convex domain D with an absorber outside. The problem is to compute the particle fission coefficient. We represent the particle transfer process as the Markov chain $x_0,\ldots,x_N$ of particle collisions with some elements in D [2,3]. Each particle collision results in scattering, absorbing, or fission, with given probabilities. We have upon the average ν new particles after each fission. The particle starts moving from the point $r_0 \in \mathrm{D}$ along the direction ω_0, i. e., from the phase point $x_0 = (r_0,\omega_0)$. On the domain boundary of D, the detector registers the particle crossings. If the particle escapes from D, then the trajectory is terminated. The problem is to compute the average number of escaping particles from domain D. The average number of the escaping particles is defined by $\varphi(x_0)$, where φ is the solution of the transfer adjoint integral equation (17.1) [1,2], where $h(r,\omega) = 1$ for $r \notin \mathrm{D}$ and $h(r,\omega) = 0$ for $r \in \mathrm{D}$. It is well known that there is such a critical value ν^* that under $\nu > \nu^*$ the medium is overcritical,, i. e., $\varphi(x_0) = +\infty$. On the other hand, under $\nu < \nu^*$ the medium is undercritical, i. e., $\varphi(x_0) < +\infty$. The simple weight estimation algorithm for the value $\varphi(x_0)$ is to simulate fission as scattering and multiply the particle weight by ν. In this case the function $\mathrm{E}\xi_x^2$ is defined by the Neumann series for the equation with the kernel that is equal to the product of the ν^2 and the kernel of initial transfer adjoint integral equation (17.1). Therefore, $\mathrm{E}\xi_x^2 < +\infty$ if $\nu^2 < \nu^*$.

It is more reasonable to use the algorithm with branching. At the point of fission we emit $[\nu]$ and $[\nu]+1$ particles with probabilities $1+[\nu]-\nu$ and $\nu-[\nu]$. According to [4] we have $\mathrm{E}\zeta_x^2 < +\infty$ for $\nu < \nu^*$. Here the average total number of branches is finite since the medium is undercritical. The final statement is that if $\varphi(x) < C < +\infty\ \forall x$, then $\mathrm{D}\zeta_x < \infty$. Note that here the condition $\rho(K) < 1$ is not obligatory.

17.3 Splitting

Here we present the approach of G. A. Mikhailov for the splitting optimization of the integral estimation, published in [2]. Suppose

$$I = \int_X \int_Y f(x, y) g(x, y) dx dy$$

is to be computed, where $f(x, y)$ is the density of the joint distribution of the random vectors ξ and η.

Following [2] let us introduce the notation $\zeta = g(\lambda, \eta)$ with j.d.f. $f(x, y)$ and

$$\mathrm{E}[\zeta|x] = \int_X f_2(y|x) g(x, y) dy, \quad f_1(x) = \int_Y f(x, y) dy,$$

where $f_1(x)$ is the density of the absolute distribution of λ; $f_2(y|x)$ is the density of the conditional distribution of η when $\lambda = x$ and $\mathrm{E}[\zeta|x]$ is the conditional mathematical expectation of the random variable ζ when $\eta = x$. Following this, the problem is reduced to the computation of the integral

$$\int_X f_1(x) \mathrm{E}[\zeta|x] dx.$$

Furthermore, the amount of the variance decreases, i.e., $\mathrm{DE}[\zeta|\lambda] \leq \mathrm{D}\zeta$ since $\mathrm{DE}[\zeta|\lambda] + \mathrm{ED}[\zeta|\lambda]$.

Let λ be distributed with density $f_1(x)$ and let $n \geq 1$ be an integer which depends on ξ. The random variables $\eta_1, \dots, \eta_{n(\xi)}$ are independent and equally conditionally distributed with density $f_2(y|x)$ under the condition $\xi = x$. G. A. Mikhailov [2] used the following estimator:

$$\zeta_n = \frac{1}{n} \sum_{i=1}^{n} g(\lambda, \eta_i), \quad \mathrm{E}\zeta_n = I, \quad \mathrm{D}\zeta_n = \mathrm{DE}[\zeta|\lambda] + \frac{\mathrm{ED}[\zeta|\lambda]}{n} = A_1 + \frac{A_2}{n}$$

and demonstrated that the use of optimal value $n^* = \sqrt{A_2 t_1 / A_1 t_2}$ minimizes the computation cost S_n^* of simulation with splitting

$$S_{n^*} = (t_1 + n t_2)(A_1 + \frac{A_2}{n^*}) \leq S_0 = (t_1 + t_2)(A_1 + A_2),$$

as compared with the computation cost S_0 for simulation without splitting ($n = 1$). Here t_1 is the computational time for the computation of one value of ξ, and t_2 is the time for the computation of one value of η under the condition $\xi = x$. For applications the variables A_1 and A_2 are estimated by the result from special a priori calculations.

Let us study the stated splitting optimization for estimating the solution of the integral equation (17.1). Direct simulation of the collision estimator with splitting the trajectory only at the second transition $x_1 \to x_2$ can be described by the following recurrence:

$$\zeta_x = h(x) + \delta_x \zeta_{x'}, \quad \zeta_{x'} = h(x') + \delta_{x'} \frac{1}{\nu} \sum_{i=1}^{\nu} \xi_{x''}^{(i)}.$$

Using step-by-step averaging [4] of ζ_x^2 and ξ_x, we derive equations for the second moments for $\mathrm{E}\zeta_x^2$ and $\mathrm{E}\xi_x^2$:

$$\begin{aligned}
\mathrm{E}\zeta_x^2 &= (A_h) + (A_\nu) \\
&= \left(h(x)[2\varphi^*(x) - h(x)] + \int_X k(x,x')h(x')[2\varphi^*(x') - h(x')]dx' \right) \\
&\quad + \left(\int_X k(x,x') \int_X k(x',x'') \frac{\nu - 1}{\nu} \varphi^{*2}(x'')dx''dx' \right. \\
&\qquad\qquad \left. + \int_X k(x,x') \int_X \frac{1}{\nu} k(x',x'') \mathrm{E}\xi_{x''}^2 dx''dx' \right)
\end{aligned}$$

$$\mathrm{E}\xi_x^2 = (A_h) + (A_d) = (A_h) + \int_X k(x,x') \int_X k(x',x'') \mathrm{E}\xi_{x''}^2 dx''dx'.$$

Let us show that the computation cost of an algorithm with splitting $S_\nu(x) = (t_1 + \nu t_2)(A_h + A_\nu)$ is always lower than the computation cost of an algorithm without splitting $S_1(x) = (t_1 + t_2)(A_h + A_d)$ where t_1 is an average computation time for $x_0 \to x_2$ and t_2 the average computation time for $x_2 \to x_N$.

It is easy to check by substitution that

$$(S_\nu(x) - S_1(x))_\nu' = \frac{t_1}{\nu^2}\left([K^*\varphi^{*2}](x) - A_d\right) + t_2\left(A_h + [K^*\varphi^{*2}](x) - \varphi^{*2}(x)\right)$$

and

$$(S_\nu(x) - S_1(x))_\nu' = 0 \ \ if \ \ \nu^* = \sqrt{\frac{t_1(A_d - [K^*\varphi^{*2}](x))}{t_2(A_h + [K^*\varphi^{*2}](x) - \varphi^{*2}(x))}}.$$

Since $(S_\nu(x) - S_1(x))_{\nu\nu}''(\nu^*) > 0$, then ν^* provides minimum to $S_\nu(x) - S_1(x)$ and

$$S_{\nu^*}(x) \geq S_1(x) \ \ \forall x \in X.$$

The gain in computation cost is the following:

$$S_1(x) - S_{\nu^*}(x) = (\nu^* - 1)^2(A_h + [K^*\varphi^{*2}](x) - \varphi^{*2}(x)).$$

Let us note that for the direct simulation of the collision estimator with splitting the trajectory only at any other transition > 2 we can perform the same operations to show that the computation cost of algorithm with splitting is always lower than the computation cost of an algorithm without splitting.

For direct simulation with splitting the trajectory only at the first transition $x_0 \to x_1$ the result is the opposite. In this case

$$\zeta_x = h(x) + \delta_x \frac{1}{\nu} \sum_{i=1}^{\nu} \xi_{x'}^{(i)},$$

$$\begin{aligned} \mathrm{E}\zeta_x^2 = {} & \int_X \frac{1}{\nu} k(x,x')\mathrm{E}\xi_{x'}^2 dx' + h(x)[2\varphi(x)^* - h(x)] \\ & + \int_X k(x,x')\frac{\nu-1}{\nu}\varphi^{*2}(x')dx'. \end{aligned}$$

In this case, by direct substitution we derive the inequality

$$\begin{aligned} S_\nu(x) - S_d(x) &= \nu t_k \mathrm{D}\zeta_x - t_k \mathrm{D}\xi_x = \\ &= t_k\Big(\nu \mathrm{D}\zeta_x - \int_X k(x,x')\mathrm{E}\xi_{x'}^2 dx' - h(x)[2\varphi^*(x) - h(x)] + \varphi^{*2}(x)\Big) \\ &= \nu t_k(\nu-1)\Big(h(x)[2\varphi^*(x) - h(x)] + \int_X k(x,x')\varphi^{*2}(x')dx' - \varphi^{*2}(x)\Big) \\ &= \nu t_k(\nu-1)\big(\varphi^{*2}(x) - [K^*\varphi^*]^2(x) + [K^*\varphi^{*2}](x) - \varphi^{*2}(x)\big) \\ &= \nu t_k(\nu-1)\big([K^*\varphi^{*2}](x) - [K^*\varphi^*]^2(x)\big) \geq 0.) \end{aligned}$$

Let us note that the last inequality is true according to the Schwarz (Cauchy–Bunyakovsky) inequality.

Acknowledgments. This work is supported by RFBR 09-01-00035 and 11-01-09340.

References

[1] G.I. Marchuk, G.A. Mikhailov, M.A. Nazaraliev et al., *Monte-carlo Method in Atmospheric Optics*, Springer-Verlag, Berlin, Heidelberg, New York, 1980.

[2] G.A. Mikhailov, *Optimization of Monte-Carlo Weighted Methods*, Springer-Verlag, Berlin, 1992.

[3] G.A. Mikhailov, and A.V. Voytishek, *Numerical Statictical Modelling. Monte-Carlo Methods*, Akademia, Moscow, 2006.

[4] G.A. Mikhailov and I.N. Medvedev, Reccurent partial in the theory of weighted Monte Carlo method, *Russ. J. Numer. Anal. Math. Modelling* **24**(3) (2009), 261–277

Author information

Ilya N. Medvedev, Institute of Computational Mathematics and Mathematical Geophysics, Siberian Branch of the Russian Academy of Sciences, Novosibirsk, Russia.
Email: `min@osmf.sscc.ru`

Chapter 18

On the Asymptotics of a Lower Bound for the Diaphony of Generalized van der Corput Sequences

Florian Pausinger and Wolfgang Ch. Schmid

Abstract. In a recent paper we constructed a lower bound for the diaphony of generalized van der Corput sequences in arbitrary base b. This bound was given in terms of basic functions, which were introduced by Faure to analyze the uniform distribution behavior of generalized van der Corput sequences. In the current chapter we apply methods of Faure to determine the asymptotics of our lower bound function in prime bases.

Keywords. Diaphony, Generalized van der Corput Sequence.

Mathematics Subject Classification 2010. 11K06, 11K38.

18.1 Introduction and Main Result

In the field of quasi-Monte Carlo methods, various notions of discrepancy are used in order to quantify the distribution behavior of an infinite sequence X. In this chapter we consider the classical diaphony $F(N, X)$, which we formally define in Section 18.2. We call an infinite sequence a low discrepancy sequence if there exists a constant c such that for all N

$$F^2(N, X) \leq c \cdot \log N.$$

As a consequence, computing the asymptotic value

$$f(X) := \limsup_{N\to\infty}(F^2(N, X)/\log N)$$

enables us to compare different sequences according to their distribution behavior. For a detailed introduction to the topic we refer to the classical book of Kuipers and Niederreiter [8] as well as to the recent book by Dick and Pillichshammer [2]. Generalized van der Corput sequences S_b^σ (generated from permutations σ or sequences of permutations Σ in base b) are examples of low-discrepancy sequences, the properties of which were studied in great detail by Faure during the past 30 years. Faure developed a method to compute asymptotic values for various notions of discrepancy, such as star discrepancy $D^*(N, X)$ or extreme discrepancy $D(N, X)$, as well as diaphony (see [1, 3]). Furthermore, he found explicit formulas for the original van der Corput sequences, which are generated from the identity permutations, and he gave examples of particularly well-distributed generalized van der Corput sequences with respect to the different discrepancies (see [1, 4]). Let

$$t^*(X) := \limsup_{N\to\infty}(D^*(N,X)/\log N), \quad t(X) := \limsup_{N\to\infty}(D(N,X)/\log N).$$

The smallest known asymptotic values have recently been improved by Ostromoukhov [9] to $t^*(S_{60}^{\Sigma'}) = 0.222223\ldots$ and $t(S_{84}^{\sigma'}) = 0.353494\ldots$, and by the authors [10] to $f(S_{57}^{\sigma'}) = 1.13794\ldots$. In 2005 Faure proved that the original van der Corput sequences have the worst distribution behavior in the class of all generalized van der Corput sequences in a certain base b [6]. Due to Faure the following bounds hold:

$$\frac{b-2}{2(b-1)\log b} \le t^*(S_b^\sigma) \le t^*(S_b^{id}), \qquad \frac{b-2}{(b-1)\log b} \le t(S_b^\sigma) \le t(S_b^{id}).$$

For exact values of $t(S_b^{id})$ and $t^*(S_b^{id})$ see [3]. Recently [11] we constructed a lower bound for the diaphony of generalized van der Corput sequences. In the present paper we show that this lower bound gives asymptotic bounds for $f(S_b^\sigma)$ like those obtained by Faure for $t^*(S_b^\sigma)$ and $t(S_b^\sigma)$. Our main result, where $\mathfrak{S}_b$ denotes the set of all permutations of $\{0, 1, \ldots, b-1\}$, is as follows:

Theorem 18.1. *For a prime base $b > 2$ let $\sigma \in \mathfrak{S}_b$, then*

$$\frac{4\pi^2(b-2)}{12(b-1)\log b} + \frac{\pi^2}{b^2(b-1)\log b} \le f(S_b^\sigma).$$

Remark 18.2. Due to Faure [1, 5], $f(S_b^\sigma)$ is bounded from above by $f(S_b^{id})$ with

$$f(S_b^{id}) = \begin{cases} \pi^2 \dfrac{b^4 + 2b^2 - 3}{48b^2 \log b}, & \text{if } b \text{ is odd,} \\[2ex] \pi^2 \dfrac{b^3 + b^2 + 4}{48(b+1)\log b}, & \text{if } b \text{ is even.} \end{cases}$$

Remark 18.3. There are other possibilities of bounding these constants (see [7, 10]). However, these other methods are all based on certain permutations in base b and thus require time consuming computations in high bases. Note that the only information we use here is the base b.

We give the most important definitions and previous the results in Section 18.2, and will prove Theorem 18.1 in Section 18.3.

18.2 Definitions and Previous Results

For an infinite sequence $X = (x_i)_{i\geq 1}$ in the half open unit interval $[0,1[$ and for $N \geq 1$ and reals $0 \leq \alpha < \beta \leq 1$ let $A([\alpha,\beta[, N, X)$ denote the number of indices $i \leq N$ for which $x_i \in [\alpha,\beta[$. Then the diaphony $F(N,X)$ of the first N points of X is defined by

$$F(N,X) := \left(2\pi^2 \int_0^1 \int_0^1 |E([\alpha,\beta[, N, X)|^2 d\alpha d\beta\right)^{1/2},$$

where $E([\alpha,\beta[, N, X) := A([\alpha,\beta[, N, X) - (\beta-\alpha)N$ denotes the discrepancy function. With this definition we follow [5]; for the classical definition of the diaphony in terms of exponential sums see Zinterhof [12].

Throughout this chapter let $b \geq 2$ and $n \geq 1$ be integers. The identity in $\mathfrak{S}_b$ is always denoted by id.

Definition 18.4. For a fixed base $b \geq 2$ and a permutation $\sigma \in \mathfrak{S}_b$ the generalized van der Corput sequence S_b^σ is defined by

$$S_b^\sigma(n) = \sum_{j=0}^{\infty} \sigma(a_j(n)) b^{-j-1},$$

where $\sum_{j=0}^{\infty} a_j(n) b^j$ is the b-adic representation of an integer $n \geq 1$.

The analysis of the diaphony of S_b^σ is based on special functions which were first introduced by Faure in [3] and which are defined as follows:

Definition 18.5. For $\sigma \in \mathfrak{S}_b$ let $Z_b^\sigma := (\sigma(0)/b, \sigma(1)/b, \ldots, \sigma(b-1)/b)$. For $h \in \{0,1,\ldots,b-1\}$ and $x \in [\frac{k-1}{b}, \frac{k}{b}[$, where $k \in \{1,\ldots,b\}$ we define

$$\varphi_{b,h}^\sigma(x) := \begin{cases} A([0,h/b[; k; Z_b^\sigma) - hx & \text{if } 0 \leq h \leq \sigma(k-1), \\ (b-h)x - A([h/b,1[; k; Z_b^\sigma) & \text{if } \sigma(k-1) < h < b. \end{cases}$$

The function $\varphi_{b,h}^\sigma$ is extended to the reals by periodicity.

In [1] Chaix and Faure introduced a new class of functions based on $\varphi_{b,h}^\sigma$:

Definition 18.6.

$$\chi_b^\sigma := \sum_{0\leq h<h'<b} (\varphi_{b,h'}^\sigma - \varphi_{b,h}^\sigma)^2.$$

Proposition 18.7 (Propriété 3.3 and Propriété 3.5 in [1]). *The following hold:*

- *the functions χ_b^σ are continuous, piecewise quadratic on the intervals $[k/b,(k+1)/b]$, and $\chi_b^\sigma(0) = \chi_b^\sigma(1)$;*

- *on each interval $[k/b, (k+1)/b]$ the parabolic arcs of χ_b^σ are translated versions of the parabola $y = b^2(b^2-1)x^2/12$.*

Remark 18.8. This means that it suffices to know the values of χ_b^σ at the b-adic positions $\frac{k}{b}$, because χ_b^σ is of the form $Ax^2 + Bx + C$, in every interval $[\frac{k}{b}, \frac{k+1}{b}]$, where A only depends on the base b.

In [11] we proved the following bounds which (due to Proposition 18.7) are enough to compute a lower bound function L_b that has the same properties as any χ_b^σ-function (see Figure 18.1):

Theorem 18.9. *For arbitrary base b and arbitrary $\sigma \in \mathfrak{S}_b$ and for each k, $1 \le k \le b$,*

$$\chi_b^\sigma\left(\frac{k}{b}\right) \ge L_b^k$$

holds, where $L_b^k := \frac{1}{2}\sum_{d=1}^{b-1}(L_b^{k,d} - \frac{k^2d^2}{b})$ and $L_b^{k,d} := (b-\gamma)\lfloor\frac{kd}{b}\rfloor^2 + \gamma\lfloor\frac{kd}{b}+1\rfloor^2$, for $kd \equiv \gamma \pmod b$, with $\gamma \in \{0, \ldots, b-1\}$.

Chaix and Faure [1] showed that the diaphony of S_b^σ can be exactly computed using χ_b^σ functions. Moreover, they developed a technique to obtain asymptotic values.

Theorem 18.10 (Théorème 4.2 in [1]). *For all $N \ge 1$, we have*

$$F^2(N, S_b^\sigma) = 4\pi^2 \sum_{j=1}^{\infty} \chi_b^\sigma(Nb^{-j})/b^2.$$

Theorem 18.11 (Théorème 4.10 in [1]). *Let*

$$\gamma_b^\sigma := \inf_{n \ge 1} \sup_{x \in [0,1]} \left(\sum_{j=1}^{n} \chi_b^\sigma(xb^{-j})/n \right),$$

then

$$f(S_b^\sigma) = \limsup_{N \to \infty}(F^2(N, S_b^\sigma)/\log N) = 4\pi^2\gamma_b^\sigma/(b^2 \log b).$$

18.3 Proof of Theorem 18.1

Let us begin this section with two technical definitions. In order to compute γ_b^σ we introduce the supporting functions

$$g_{b,n}^\sigma := \frac{1}{n}\sum_{j=1}^{n} \chi_b^\sigma(xb^{-j})/n.$$

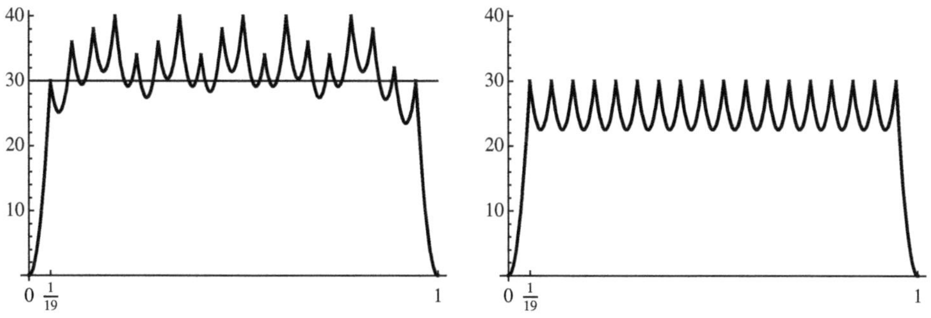

Figure 18.1. Left: a χ_{19}^{σ}-function of a sequence in base 19. Right: lower bound function L_{19}.

Following [1, 3] we examine intervals $I_h^n := [h/b^n, (h+1)/b^n]$, with $h \in \{0, \ldots, b^n - 1\}$. We call the interval I_h^n *dominated* if there exists a set J of integers with $h \notin J$ such that $g_{b,n}^{\sigma}(x) \leq \max_{j \in J} g_{b,n}^{\sigma}((x + (j-h))/b^n)$, for all $x \in I_h^n$. Otherwise the interval is called *dominant*. It is enough to consider dominant intervals of the functions $g_{b,n}^{\sigma}$ to determine the supremum of $g_{b,n+1}^{\sigma}$. Since our lower bound functions show a very regular pattern concerning dominant intervals (see Lemma 18.12) we can explicitly compute the supremum of $g_{b,n}^{\sigma}$ and determine its asymptotic value as $n \to \infty$.

The following lemma proves that for any lower bound function in prime base b the interval $[1/b, 2/b]$ is dominant.

Lemma 18.12. *For arbitrary base b and for k, with* $1 \leq k \leq b-1$,

$$L_b^k = L_b^1 = \frac{1}{12}(b+1)(b-1)$$

holds, if k and b are coprime.

Proof. We have $L_b^k = \frac{1}{2}\sum_{d=1}^{b-1}((b-\gamma)\lfloor\frac{kd}{b}\rfloor^2 + \gamma\lfloor\frac{kd}{b} + 1\rfloor^2 - \frac{k^2d^2}{b})$ for $kd \equiv \gamma \pmod{b}$, with $\gamma := \gamma(k,d,b) \in \{0, \ldots, b-1\}$. First note that $\lfloor\frac{kd}{b}\rfloor = \frac{kd-\gamma}{b}$ since $\frac{kd-\gamma}{b} \in \mathbb{Z}$ and $\frac{\gamma}{b} \in \{0, \frac{1}{b}, \ldots, \frac{b-1}{b}\}$. Then it follows by easy calculation that

$$\begin{aligned} L_b^k &= \frac{1}{2}\sum_{d=1}^{b-1}\left((b-\gamma)\left(\frac{kd-\gamma}{b}\right)^2 + \gamma\left(\frac{kd-\gamma}{b}+1\right)^2 - b\left(\frac{kd}{b}\right)^2\right) \\ &= \frac{1}{2}\sum_{d=1}^{b-1}\left(\gamma - \frac{\gamma^2}{b}\right). \end{aligned}$$

Note that for k with $\gcd(k,b) = 1$ we have $\{k \cdot 1, k \cdot 2, \ldots, k \cdot (b-1)\} \pmod{b} = \{1, 2, \ldots, b-1\}$ and therefore

$$L_b^k = \frac{1}{2} \sum_{d=1}^{b-1} \left(d - \frac{d^2}{b} \right) = \frac{1}{12}(b+1)(b-1). \qquad \square$$

Remark 18.13. Numerical investigations show that for arbitrary base b and for k, with $1 \leq k \leq b-1$, we always have $L_b^k \leq L_b^1$, with equality only if k and b are coprime.

Now we can compute the coefficients of L_b in the interval $[1/b, 2/b]$. Note that L_b is (due to Proposition 18.7) quadratic in each interval $[k/b, (k+1)/b]$ and completely defined by the set $\{L_b^k : 1 \leq k \leq b\}$. Hence we need to solve the linear system

$$L_b(1/b) = L_b^1, \quad L_b(2/b) = L_b^2,$$

which gives $B = 1/4(b-b^3), C = 1/4(b^2-1)$. Consequently, due to Theorems 18.9, and 18.11 and Lemma 18.12 we get for arbitrary b and σ

$$\begin{aligned} \gamma_b^\sigma &\geq \inf_{n \geq 1} \sup_{x \in [0,1]} \left(\sum_{j=1}^{n} L_b(xb^{-j})/n \right) \\ &= \inf_{n \geq 1} \frac{1}{n} \left(\frac{b^2(b^2-1)}{12} \left(\bar{x}_0^2 + \sum_{i=1}^{n-1} b^i \bar{x}_i^2 \right) + B \sum_{i=1}^{n-1} b^i \bar{x}_i + nC \right), \end{aligned}$$

where $\bar{x}_i := \sum_{j=i+1}^{n} (1/b^j)$.

For $n \to \infty$ and after some easy calculations we get

$$\begin{aligned} \gamma_b^\sigma &\geq \frac{b^6+b^2}{12(b-1)^3(b+1)} - \frac{b^4}{6(b-1)^3(b+1)} - \frac{b^4-b^3-b^2+b}{4(b-1)^2} + \frac{b^2-1}{4} \\ &= \frac{(b-2)b^2+3}{12(b-1)} = \frac{(b-2)b^2}{12(b-1)} + \frac{1}{4(b-1)}. \end{aligned}$$

Thus we can finish the proof of Theorem 18.1 as follows:

$$\begin{aligned} f(S_b^\sigma) &= \frac{4\pi^2 \gamma_b^\sigma}{b^2 \log b} \\ &\geq \frac{4\pi^2}{b^2 \log b} \left(\frac{(b-2)b^2}{12(b-1)} + \frac{1}{4(b-1)} \right) \\ &= \frac{4\pi^2(b-2)}{12(b-1)\log b} + \frac{\pi^2}{b^2(b-1)\log b}. \end{aligned} \qquad \square$$

Acknowledgments. This work is supported by the Austrian Science Foundation (FWF), Project S9609, which is part of the Austrian National Research Network "Analytic Combinatorics and Probabilistic Number Theory".

References

[1] H. Chaix and H. Faure, Discrépance et diaphonie en dimension un, *Acta Arith.* **63** (1993) 103–141.

[2] J. Dick and F. Pillichshammer, Digital Nets and Sequences. Cambridge University Press, Cambridge, 2010.

[3] H. Faure, Discrépance de suites associées à un système de numération (en dimension un). *Bull. Soc. Math. France* **109** (1981) 143–182.

[4] H. Faure, Good Permutations for Extreme Discrepancy, *Journal of Number Theory* **42** (1992), 47–56.

[5] H. Faure, Discrepancy and diaphony of digital (0, 1)-sequences in prime base, *Acta Arith.* **117** (2005), 125–148.

[6] H. Faure, Irregularities of Distribution of Digital (0, 1)-Sequences in Prime Base, *Electronic Journal of Combinatorial Number Theory* **5**(3), (2005) 1–12.

[7] H. Faure, Improvements on low discrepancy one-dimensional sequences and two-dimensional point sets, in: A. Keller et al. (eds.) *Monte Carlo and Quasi-Monte Carlo Methods 2006*, pp. 327-341, Springer-Verlag, Berlin Heidelberg, 2008.

[8] L. Kuipers and H. Niederreiter, *Uniform distribution of sequences*, Wiley, New York, 1974.

[9] V. Ostromoukhov, Recent Progress in Improvement of Extreme Discrepancy and Star Discrepancy of One-Dimensional Sequences, in: P. L'Ecuyer and A.B. Owen (eds.), *Monte Carlo and Quasi-Monte Carlo Methods 2008*, pp. 561–572, Springer-Verlag, Berlin, 2009.

[10] F. Pausinger and W. Ch. Schmid, A good permutation for one-dimensional diaphony, *Monte Carlo Methods Appl.* **16** (2010), 307–322.

[11] F. Pausinger and W. Ch. Schmid, A lower bound for the diaphony of generalised van der Corput sequences in arbitrary base b, *Unif. Distrib. Theory* **6**(2) (2011), 31–46.

[12] P. Zinterhof, Über einige Abschätzungen bei der Approximation von Funktionen mit Gleichverteilungsmethoden. *Österr. Akad. Wiss. SB II* **185** (1976), 121–132.

Author information

Florian Pausinger, Institute of Science and Technology Austria (IST Austria), Klosterneuburg, Austria.
Email: `florian.pausinger@ist.ac.at`

Wolfgang Ch. Schmid, Department of Mathematics, University of Salzburg, Salzburg, Austria.
Email: `wolfgang.schmid@sbg.ac.at`

Chapter 19

Group Object Tracking with a Sequential Monte Carlo Method Based on a Parameterized Likelihood Function

Nikolay Petrov, Lyudmila Mihaylova, Amadou Gning, and Donka Angelova

Abstract. Group objects are characterized by multiple measurements originating from different locations of the targets constituting the group. This chapter presents a novel sequential Monte Carlo approach for tracking groups with a large number of components, applicable to various nonlinear problems. The novelty in this work is in the derivation of the likelihood function for nonlinear measurement functions, with sets of measurements belonging to a bounded spatial region. Simulation results are presented when a group of 50 objects is surrounded by a circular region. Estimation results are given for the group object center and extent.

Keywords. Sequential Monte Carlo Methods, Measurement Uncertainty, Nonlinear Estimation, Group Object Tracking.

Mathematics Subject Classification 2010. 62Lxx, 62Mxx, 93E10, 60J05, 60J20, 90B50, 62Gxx, 65C05, 93Exx.

19.1 Motivation

Group object tracking is concerned with finding the patterns of behavior and tracks of a whole group instead of each object separately. The group of objects are generating multiple measurements the origin of which is unknown. Different methods have been proposed in the literature for solving the problem of group object tracking, e. g., [7, 8]. Several approaches consist of estimating the extent of the group, e. g., with random matrices, as suggested in [2, 7]. Other related works are [3, 4, 6, 9–12].

In general the measurement uncertainty can pertain to a hypercube or to another spatial shape. In our approach, we consider the general case with a nonlinear measurement equation and measurements. The main contributions of this work is in the derived likelihood function based on a parameterized shape and in the developed sequential Monte Carlo filter for group objects. Then we propagate this uncertainty in spatial measurement through the Bayesian estimation framework.

In this work, inspired by ideas from [5, 6], we developed a novel approach for tracking groups of objects consisting of a large number of components. The remaining part of the chapter is structured as follows. Section 19.2 formulates the problem,

Section 19.3 derives the measurement likelihood function, Section 19.4 presents simulation results about the performance of the proposed approach, and finally Section 19.5 summarizes the results.

19.2 Group Object Tracking within the Sequential Monte Carlo Framework

The problem of state estimation of groups of objects is considered within the sequential Monte Carlo framework. Each group, $g = 1, \ldots, n_g$, is comprized of n_T^g targets. The group state vector is defined as $\boldsymbol{X}_k^g = (\boldsymbol{x}'_{g,k}, \boldsymbol{G}'_k)'$, where $\boldsymbol{x}_{g,k}$ is the vector containing the position and speed coordinates of the center of the region surrounding the group g, at time k, $\boldsymbol{G}_k$ is a vector characterizing the group, and $'$ is the transpose operator. The groups of objects give rise to a set of measurements $\boldsymbol{Z}_k = \{z_{m,k}\}_{m=1}^{M_k}$. The goal is to estimate $\boldsymbol{X}_k^g$ given the measurements $\boldsymbol{Z}_{1:k} = \{\boldsymbol{Z}_1, \ldots, \boldsymbol{Z}_k\}$, collected up to time k. Within the sequential Bayesian framework this goal can be achieved by constructing the group state probability density function (pdf) based on the incoming sensor data. According to Bayes' rule the filtering pdf $p(\boldsymbol{X}_k^g|\boldsymbol{Z}_{1:k})$ of the state vector $\boldsymbol{X}_k^g$, given a sequence of sensor measurements $\boldsymbol{Z}_{1:k}$, can be written as

$$p(\boldsymbol{X}_k^g|\boldsymbol{Z}_{1:k}) = \frac{p(\boldsymbol{Z}_k|\boldsymbol{X}_k^g)p(\boldsymbol{X}_k^g|\boldsymbol{Z}_{1:k-1})}{p(\boldsymbol{Z}_k|\boldsymbol{Z}_{1:k-1})}, \tag{19.1}$$

where $p(\boldsymbol{Z}_k|\boldsymbol{Z}_{1:k-1})$ is the normalizing constant. The state *predictive* distribution is given by the equation

$$p(\boldsymbol{X}_k^g|\boldsymbol{Z}_{1:k-1}) = \int p(\boldsymbol{X}_k^g|\boldsymbol{X}_{k-1}^g)p(\boldsymbol{X}_{k-1}^g|\boldsymbol{Z}_{1:k-1})d\boldsymbol{X}_{k-1}^g. \tag{19.2}$$

The evaluation of the right-hand side of (19.1) involves integration which can be performed within the sequential Monte Carlo framework (known also as particle filtering) [1] by approximating the posterior pdf $p(\boldsymbol{X}_k^g|\boldsymbol{Z}_{1:k})$ with a set of particles $\boldsymbol{X}_k^{g(i)}$, $i = 1, \ldots, N$ and their corresponding weights $w_k^{(i)}$. Then the posterior density function can be written as follows:

$$p(\boldsymbol{X}_k^g|\boldsymbol{Z}_{1:k}) \approx \sum_{i=1}^{N} w_k^{(i)}\delta(\boldsymbol{X}_k^g - \boldsymbol{X}_k^{g(i)}), \tag{19.3}$$

where $\delta(\cdot)$ is the Dirac delta function, and the weights are normalized such that $\sum_i w_k^{(i)} = 1$.

Each pair $\{\boldsymbol{X}_k^{g(i)}, w_k^{(i)}\}$ characterises the belief that the group g is in state $\boldsymbol{X}_k^{g(i)}$. Note that, for example in the case of a circle, we assume that $\boldsymbol{X}_k^{g(i)} = (x_{c,k}, y_{c,k}, \dot{x}_{c,k}, \dot{y}_{c,k}, r_k)'$, where $x_{c,k}, y_{c,k}$ are the center position coordinates, $\dot{x}_{c,k}$,

$\dot{y}_{c,k}$ are the respective speed components, and r_k is the radius. *In a general case a particle* $\boldsymbol{X}_k^{g(i)}$ *characterizes a spatial shape.* An estimate of the variable of interest is obtained by the weighted sum of the particles. Two major stages can be distinguished: *prediction* and *update*. During prediction, each particle is modified according to the state evolution model, including the addition of random noise. In the measurement update stage, the weights are reevaluated using a likelihood term. The residual resampling [13] is added to prevent sample impoverishment.

19.3 Measurement Likelihood for Group Object Tracking

In the context of group object tracking, the prediction step is generally well studied for various classes of interval and spatial representations of uncertainties. Ellipsoids, spheres and polytope families can be easily propagated when the system dynamics and sensor models are linear.

The update step with the likelihood calculation is less studied or often studied with a restriction to a particular class of sets or a particular type of measurement models. The aim of this paper is to derive general likelihood calculation procedures based on Monte Carlo methods without a restriction on the type of regions of interest or the type of measurement available. For this purpose, the main idea of this chapter is to introduce a sampling step for regions of interest in the groups regions. This sampling aims to represent the probability of a given point, in the state space, to be the origin of a measurement. This section describes in detail the newly proposed method. As an illustration, the case of a group object with circular form is considered.

The sensor state of sensor s at time k is denoted by $\boldsymbol{x}_{s,k}$. Assume that at each measurement time k the groups generate a matrix $\boldsymbol{Z}_k$ of $M_k = \sum_{g=1}^{G}\sum_{t=1}^{n_{Tg}} n_T^g M_{t,k}$ measurement vectors. The number of measurements $M_{k,t}$ originating from the target t is considered to be Poisson-distributed random variable with mean value of μ_t, i. e., $M_{t,k} \sim Poisson(\mu_t)$. The measurements originating from a target t, having a state vector $\boldsymbol{x}_{t,k}^g$ are assumed to be conditionally independent, i. e.,

$$p(\boldsymbol{Z}_k|\boldsymbol{x}_{t,k}) = \prod_{m=1}^{M_{t,k}} p(\boldsymbol{z}_{m,k}|\boldsymbol{x}_{t,k}^g). \tag{19.4}$$

For each target t, the generation of measurements over the observation space at time step k is modeled as a nonhomogeneos Poisson point process (PPP). Let $\lambda_t(\boldsymbol{z}_{m,k}|\boldsymbol{x}_{t,k}, \boldsymbol{G}_k)$ be the intensity of the PPP. The pdf of each measurement $\boldsymbol{z}_m$, produced in a region A of the observation space (sensor field-of-view) can be expressed by $p(\boldsymbol{z}_{m,k}|\boldsymbol{x}_{t,k}^g, \boldsymbol{G}_k) = \lambda_t(\boldsymbol{z}_{m,k}|\boldsymbol{x}_{t,k}^g, \boldsymbol{G}_k)/\mu_t(A)$, where $\mu_t(A)$ is the expected number of measurements falling in A. We assume that $\mu_t(A)$ is known. It is assumed also that the measurements received over the observation volume result from the superposition of n_T^g conditionally independent nonhomogeneous PPPs. Since the superposition of such independent Poisson processes is a Poisson process with a parameter that is $\sum_{t=1}^{n_t^g} \lambda_t$, where the expected total number of measurements, received in the time instant k, is given by $\mu(A) = \sum_{t=1}^{n_t^g} \mu_t(A)$. Similarly to [5], the joint likelihood of the number n_T^g of measurement falling in the observation volume A is given by

$$p(\boldsymbol{Z}_k|\boldsymbol{X}_k^g) \approx \prod_{m=1}^{M_k} \sum_{t=1}^{n_T^g} \mu_t(A) p(\boldsymbol{z}_{m,k}|\boldsymbol{x}_{t,k}^g). \tag{19.5}$$

Note that in this work the state vector $\boldsymbol{x}_{t,k}^g$ of each target is not estimated. Instead the proposed technique estimates the center of the region and its extent.

19.3.1 Introduction of the Notion of the Visible Surface

The measurement likelihood $p(\boldsymbol{z}_{m,k}|\boldsymbol{X}_k^g)$ of the group target can be represented with the relation

$$p(\boldsymbol{z}_{m,k}|\boldsymbol{X}_k^g) = \int_{\mathbb{R}^{n_\mathrm{v}}} p(\boldsymbol{z}_{m,k}|\boldsymbol{V}_k^g) p(\boldsymbol{V}_k^g|\boldsymbol{X}_k^g) d\boldsymbol{V}_k^g, \tag{19.6}$$

where $\boldsymbol{V}_k^g \in \mathbb{R}^{n_v}$ denotes a source of measurement in the state space. In practice, these visible sources $\boldsymbol{V}_k^g$ depend on the group state vector $\boldsymbol{X}_k^g$ and on the sensor state (e. g., the sensor position and angle of view). The pdf $p(\boldsymbol{V}_k^g|\boldsymbol{X}_k^g)$ represents the probability of a point in the state space to be a source of measurement given the group $\boldsymbol{X}_k^g$. The surface of the group with state $\boldsymbol{X}_k^g$ visible from the sensor with a state vector $\boldsymbol{x}_{s,k}$ is denoted by $\mathcal{V}_k(\boldsymbol{X}_k^g, \boldsymbol{x}_{s,k})$.

We assume that the sources of the measurements at time step k, given the target state and the sensor prior state are uniformly distributed along the region $\mathcal{V}_k(\boldsymbol{X}_k^g, \boldsymbol{x}_{s,k})$, visible from the sensor $\boldsymbol{x}_{s,k}$, i. e.,

$$p(\boldsymbol{V}_k^g|\boldsymbol{X}_k^g) = p(\boldsymbol{V}_k^g|\boldsymbol{X}_k^g, \boldsymbol{x}_{s,k}) = \mathcal{U}_{\mathcal{V}_k(\boldsymbol{X}_k^g, \boldsymbol{x}_{s,k})}(\boldsymbol{V}_k^g), \tag{19.7}$$

where $\mathcal{U}_{\mathcal{V}_k(\boldsymbol{X}_k^g, \boldsymbol{x}_{s,k})}(\cdot)$ represents the uniform pdf in $\mathcal{V}_k(\boldsymbol{X}_k^g, \boldsymbol{x}_{s,k})$.

Inserting (19.7) into (19.6) gives

$$p(\boldsymbol{z}_{m,k}|\boldsymbol{X}_k^g) = \int_{\mathbb{R}^{n_\mathrm{v}}} p(\boldsymbol{z}_{m,k}|\boldsymbol{V}_k^g) \mathcal{U}_{\mathcal{V}_k(\boldsymbol{X}_k^g, \boldsymbol{x}_{s,k})}(\boldsymbol{V}_k^g) d\boldsymbol{V}_k^g. \tag{19.8}$$

19.3.2 Parametrization of the Visible Surface

The likelihood (19.6) can be calculated using the following Monte Carlo integration:

$$\begin{aligned} p(z_{m,k}|\boldsymbol{X}^{g(i)}_{k|k-1}) &= \int_{\mathbb{R}^{n_x}} p(z_{m,k}|\boldsymbol{V}^g_k)p(\boldsymbol{V}^g_k|\boldsymbol{X}^{g(i)}_{k|k-1})d\boldsymbol{V}^g_k \\ &\approx \sum_{\ell=1}^{S} p(z_{m,k}|\boldsymbol{V}^{g(\ell)}_k)p(\boldsymbol{V}^{g(\ell)}_k|\boldsymbol{X}^{g(i)}_{k|k-1}). \end{aligned} \tag{19.9}$$

After the prediction step, a set of weighted particles $\{(\boldsymbol{X}^{g(i)}_{k|k-1}, w^{(i)}_{k|k-1})\}^N_{i=1}$ is available. First, the visible area $\mathcal{V}_k(\boldsymbol{X}^{g(i)}_{k|k-1}, \boldsymbol{x}_{s,k})$ is determined. Then for each particle $\boldsymbol{X}^{g(i)}_{k|k-1}$ the likelihood function $p(\boldsymbol{V}^{g(\ell)}_k|\boldsymbol{X}^{g(i)}_{k|k-1})$ is defined from $\mathcal{V}_k(\boldsymbol{X}^{g(i)}_{k|k-1}, \boldsymbol{x}_{s,k})$, with $\ell = 1, \ldots, S$, where $S = QN$ and Q is the number of point samples generated uniformly inside each of the N circle-particles, characterizing the group extent.

The term $p(z_{m,k}|\boldsymbol{V}^{g(\ell)}_k)$ in (19.9) can be easily calculated in a classical way, depending on the problem, for example using a Gaussian measurement noise distribution:

$$p(z_{m,k}|\boldsymbol{V}^{g(\ell)}_k) = \frac{1}{\sqrt{2\pi||\boldsymbol{R}||}} e^{-\frac{(z_{m,k}-z^{(\ell)}_k)\boldsymbol{R}^{-1}(z_{m,k}-z^{(\ell)}_k)^T}{2}}, \tag{19.10}$$

where $\boldsymbol{R}$ is a known measurement noise covariance matrix.

A simple expression of $p(\boldsymbol{V}^{g(\ell)}_k|\boldsymbol{X}^{g(i)}_{k|k-1})$, in the case of circular group shape, can be

$$p\left(\boldsymbol{V}^{g(\ell)}_k|\boldsymbol{X}^{g(i)}_{k|k-1}\right) = \mathcal{U}_{\mathcal{C}(x^{(i)}_{c,k}, y^{(i)}_{c,k}, r^{(i)}_{k|k-1})}\left(x^{(\ell)}_k, y^{(\ell)}_k\right), \tag{19.11}$$

19.4 Performance Evaluation

An example with one group that consists of 50 targets is considered for the duration of 200 time steps. The scenarios are repeated for 20 Monte Carlo iterations. We are estimating the vector $\boldsymbol{X}^g_k = \left(x_{c,k}, y_{c,k}, \dot{x}_{c,k}, \dot{y}_{c,k}, r_k\right)'$, with $n_g = 1$, characterizing the group. The initial coordinates of the targets are uniformly distributed in a square 50 m × 50 m. A constant velocity model is considered for the target evolution model. The initial position is defined by the mass center of the actual targets positions with added Gaussian noise with standard deviation of 5 m for both coordinates $x_{c,k=0}$ and $y_{c,k=0}$. The velocity is $\dot{x}_{c,k=0} = \mathcal{N}(100, 2^2)$ m/s and $\dot{x}_{c,k=0} = \mathcal{N}(100, 2^2)$ m/s, respectively. The system dynamic noise is characterised by $\sigma_x = \sigma_y = 0.1$.

Measurements consisting of range and bearing are considered. They are collected by sensors randomly positioned the and covering the target trajectory. The number of the measurements originating from a target t is Poisson distributed random variable

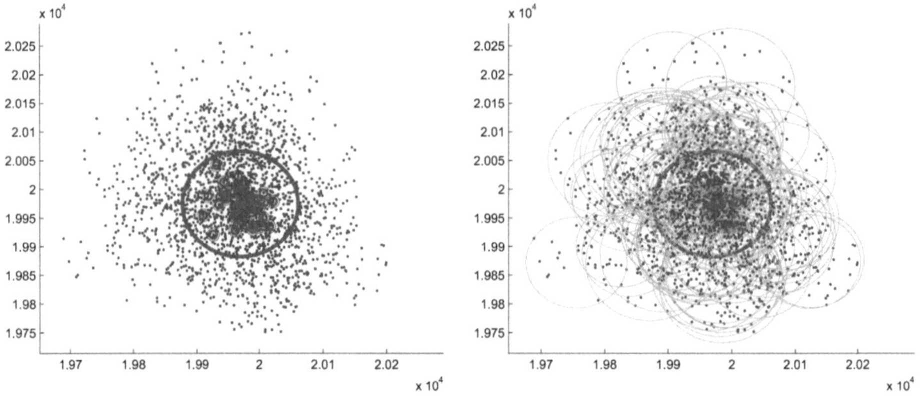

(a) (b)

Figure 19.1. Visualization of the considered simulation scenario. (a) Samples for the Monte Carlo integration of (19.9) - -the dots, estimated extent – the big circle, targets – the small double circles; (b) Particles – light circles within the big circle.

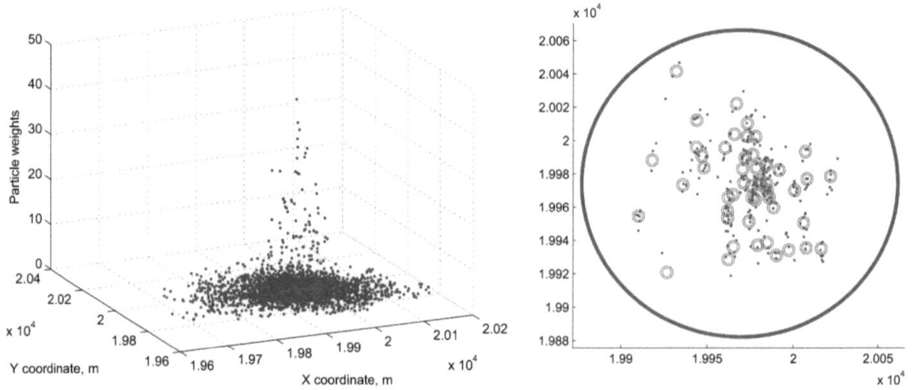

(a) (b)

Figure 19.2. Visualization of the considered simulation scenario. (a) Samples for the Monte Carlo integration of (19.9) with their weights – the dots. (b) Targets – small double circles, measurements – the dots, estimated group extent – the large circle.

with $\mu_t = 5$. The performance of the algorithm has been evaluated with a variable number of particles N from 20 to 300. For each of these circle-particles the integration of (19.9) is carried out with $Q = 5$ uniformly distributed point samples.

Figures 19.1a and b and 19.2a and b represent different plots of the same time step for a scenario similar to the one described above, with $N = 100$ circle particles and $Q = 20$ point samples. The visualization of the sampling is represented with the dots on in Figure 19.1a and b, while the particles are plotted with the light circles in Figure 19.1b. The point samples with their weights are shown in Figure 19.2a. In Figure 19.2b

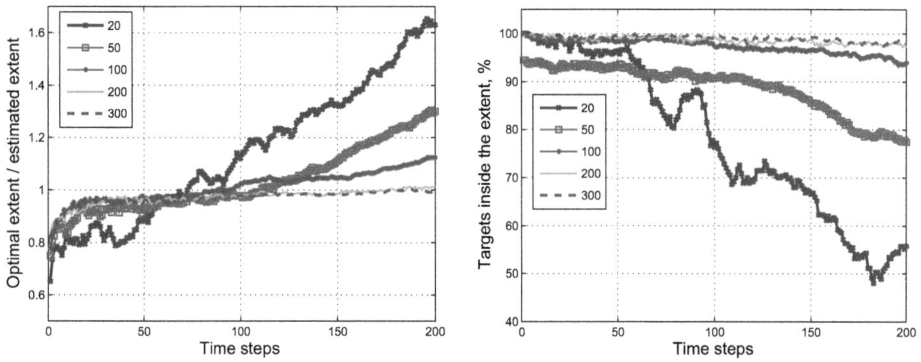

(a) (b)

Figure 19.3. Simulation results for different number of particles. (a) Ratio between the optimal and the estimated group extent; (b) Percentage of targets within the estimated extent.

the real targets can be seen as small double circles, the measurements are represented as dots, the estimated group extent is given as the big circle.

For the purpose of comparison, we define an optimal extent, which is the smallest circle surrounding all the targets within the group. Next, based on the proposed algorithm the group extent is estimated. We form the ratio between these two extents (estimated over optimal). This ratio as well as the number of the targets belonging to the estimated extent is shown in Figure 19.3. The first plot shows a convergence to an idealistic circle. In the second plot we see that the percentage of the targets contained within the estimated circle is close to 100 % for a sufficient number of particles, $N = 200$ or more.

19.5 Conclusions

This chapter presents a sequential Monte Carlo approach for group object tracking. The novelty of this work consists in the proposed algorithm for the likelihood estimation. The performance of the algorithm is validated over an example with a group having 50 components. We show that the algorithm successfully evaluates the center and the group extent based on measurements with an unknown origin coordinates. The algorithm is scalable and applicable to a large number of components (e. g., hundreds or thousands).

Acknowledgments. We acknowledge the support from the [European Community's] Seventh Framework Programme [FP7/2007–2013] under grant agreement No. 238710 (Monte Carlo based Innovative Management and Processing for an Unrivalled Leap in Sensor Exploitation), the EU COST action TU0702, and the Bulgarian National Science Fund, grant DTK 02/28/2009.

References

[1] M. Arulampalam, S. Maskell, N. Gordon, and T. Clapp, A Tutorial on Particle Filters for Online Nonlinear/Non-Gaussian Bayesian Tracking, *IEEE Trans. on Signal Proc.* **50** (2002), 174–188.

[2] M. Baum, M. Feldmann, D. Fränken, U. D. Hanebeck, and W. Koch, Extended Object and Group Tracking: A Comparison of Random Matrices and Random Hypersurface Models, in: *GI Jahrestagung (2)*, pp. 904–906, 2010.

[3] M. Baum and U. D. Hanebeck, Extended Object Tracking based on Combined Set-Theoretic and Stochastic Fusion, in: *Proc. of the International Conf. on Information Fusion*, Ajman, United Arab Emirates, December 2009.

[4] M. Baum and U. D. Hanebeck, Random Hypersurface Models for extended object tracking, in: *Proc. of the IEEE International Symp. on Signal Processing and Information Technology (ISSPIT)*, pp. 178–183, 2009.

[5] K. Gilholm, S. Godsill, S. Maskell, and D. Salmond, Poisson Models for Extended Target and Group Tracking, in: *Proceedings of SPIE 5913*, 2005.

[6] K. Gilholm, S. Godsill, S. Maskell, and D. Salmond, Poisson Models for Extended Target and Group Tracking, in: *Proc. SPIE 5913, Signal and Data Processing of Small Targets 2005*, 59130R, September 15, 2005, doi:10.1117/12.618730.

[7] J. Koch, Bayesian approach to extended object and cluster tracking using random matrices, *Aerospace and Electronic Systems, IEEE Transactions* **44** (2008), 1042–1059.

[8] W. Koch and M. Feldmann, Cluster tracking under kinematical constraints using random matrices, *Robotics and Autonomous Systems* **57** (2009), 296 – 309.

[9] B. Noack, V. Klumpp, N. Petkov, and U. D. Hannebeck, Bounding Linearisation Errors with Sets of Densities in Approximate Kalman Filtering, in: *Proc. of the 13th International Conf. on Information Fusion*, ISIF, Edinburgh, UK, 2010.

[10] F. Sawo, D. Brunn, and U. D. Hanebeck, Parameterized Joint Densities with Gaussian and Gaussian Mixture Marginals, in: *Proceedings of the 9th International Conf. on Information Fusion*, Florence, Italy, 2006.

[11] O. C. Schrempf and U. D. Hanebeck, Recursive Prediction of Stochastic Nonlinear Systems Based on Dirac Mixture Approximations, in: *Proceedings of the American Control Conference 2007. ACC '07*, July 2007, pp. 1768–1774, 2007, DOI 10.1109/ACC.2007.4282938

[12] O. C. Schrempf and U. D. Hanebeck, A state estimator for nonlinear stochastic systems based on Dirac mixture approximations, in: *ICINCO-SPSMC*, pp. 54–61, 2007.

[13] E. Wan and R. van der Merwe, The Unscented Kalman Filter, in: S. Havkin (ed.), *Kalman Filtering and Neural Networks*, Chap. 7, pp. 221–280, Wiley Publishing, New York, 2001.

Author information

Nikolay Petrov, School of Computing and Communication Systems, Lancaster University, Lancaster, UK.
Email: n.petrov@lancaster.ac.uk

Lyudmila Mihaylova, School of Computing and Communication Systems, Lancaster University, Lancaster, UK.
Email: mila.mihaylova@lancaster.ac.uk

Amadou Gning, School of Computing and Communication Systems, Lancaster University, Lancaster, UK.
Email: e.gning@lancaster.ac.uk

Donka Angelova, Institute of Information and Communication Technologies, Bulgarian Academy of Sciences, Sofia, Bulgaria.
Email: donka@bas.bg

Monte Carlo Methods and Applications, 181–191

Chapter 20

The Template Design Problem: A Perspective with Metaheuristics

David Rodríguez Rueda, Carlos Cotta, and Antonio J. Fernández-Leiva

Abstract. This chapter deals with the template design problem, a hard constrained combinatorial problem with multiple applications. This problem is formulated here as a two-level combinatorial optimization problem whose solution is integer matrices. In the higher level, a metaheuristic tackles the design of a collection of templates containing multiple instances of a set of components to be produced; in the lower level an integer linear programming solver is used to determine the optimal number of times each template has to be pressed in order to fulfill production requirements as closely as possible. Three metaheuristics, i. e., hill climbing, tabu search, and genetic algorithms, have been considered in the higher level, and LPSolve, a simplex-based software for linear and integer programming problems, in the lower level. An empirical evaluation on three scenarios of increasing complexity has been performed, indicating the better performance of genetic algorithms. These results are comparable to those shown by sequential integer linear programming (ILP) models, and hint the possibility of hybrid approaches.

Keywords. Template Design Problem, Optimization Problem, Metaheuristics, ILP Model.

Mathematics Subject Classification 2010. 90b18, 90B50, 68M10, 90C29.

20.1 Introduction

Many problems in manufacturing are related to the reduction of the waste of the raw materials used in the production process. This is precisely the case of the template design problem (TDP) considered in this article. The TDP arises in industrial settings in which variations of certain products must be produced, each of them requiring a particular packaging (typically with different printing patterns). The production of these packages must be accomplished with minimal use of cardboard (or any other used raw materials). Appropriate templates for printing these packages must then be designed, hence leading to the TDP; see Section 20.2 for a more detailed description of the problem.

Several methods have been proposed in the literature to attack this problem, which turns out to be an extremely difficult one. Our proposal is based on the use of metaheuristics, which to the best of our knowledge is a novel approach for this problem. More specifically, we consider both local search methods, i. e., a steepest descent hill climbing (HC) algorithm and tabu search (TS), and population-based techniques, i. e., a genetic algorithm (GA) (see Section 20.3). These techniques are deployed on the TDP with the aim of finding adequate template designs. Such designs must be evaluated in terms of their optimal usage. For this purpose, an ILP model shown by Proll and Smith [7] is used and solved with an exact method, hence leading to a two-level optimization approach. As will be shown in Section 20.4, the resulting techniques can effectively provide high quality solutions when applied to several instances taken from the literature.

20.2 The Template Design Problem

The template design problem was first described by Proll and Smith [7] who observed this problem at a local color printing firm. Roughly speaking, we can assume a certain product is manufactured with several variations (e. g., cereal flakes of different flavors), each one requiring a similar, but different packaging. In order to produce the packaging, a printing machine is used. This machine is configured with a certain template, which is subsequently pressed on sheets of raw material (e. g., cardboard). Given the large number of items required, a template comprises several slots, each of them filled with a certain variation of the product, which are printed on each pressing. In addition, there can be more than one such template. This means that the problem is twofold: i) determine the design of each template, namely which variations are included in each slot, and ii) determine the optimal usage of these templates. The last point requires optimization of a certain criterion, e. g., minimization of the manufacturing time (i. e., minimizing the number of pressings) or minimization of waste, given the known demands of each variation. We consider here this latter criterion, i. e., optimization of the use of raw materials.

In order to solve subproblem ii) above, an ILP formulation provided by Proll and Smith is used. Let V be the number of variations to be produced; let there be T templates $T_1, \dots, T_n$, each of them with S slots. Notice that the particular slot in which a variation is placed is irrelevant: it only matters how many instances of a certain variation are contained in a given template. Thus, let p_{ij} be the number of instances of variation i in template T_j. Now, let Q_i be the demand for variation i (deterministic and known; see [6] for an approach under uncertainty), and let us assume that we have production tolerances l_i, u_i for each variation, i. e., we can afford up to a certain level of underproduction $Q_i l_i$ or overproduction $Q_i u_i$ for each variation i. Then, the resulting problem is

$$\varphi = \min \sum_{i=1}^{V} (U_i + O_i) \tag{20.1}$$

subject to

$$\sum_{j=1...T} p_{ij} R_j + U_i - O_i = Q_i \qquad 1 \leqslant i \leqslant V,$$

$$\sum_{j=1...T} p_{ij} R_j \geqslant (1 - l_i) Q_i \qquad 1 \leqslant i \leqslant V,$$

$$\sum_{j=1...T} p_{ij} R_j \leqslant (1 + u_i) Q_i \qquad 1 \leqslant i \leqslant V,$$

$$U_i, O_i, R_j \geqslant 0, \quad 1 \leqslant i \leqslant V \quad 1 \leqslant j \leqslant T, \tag{20.2}$$

where U_i and O_i are slack variables representing respectively the underproduction and overproduction of variation i, and R_j is the number of times template j is punched (pressed). This ILP formulation can be solved using an appropriate exact technique such as branch-and-bound or branch-and-cut. There still remains the problem of determining the design of the templates used as input in this ILP formulation. This is precisely where metaheuristics come into play.

20.3 Solving the TDP under Deterministic Demand

The TDP exhibits a clear combinatorial structure and can be readily transformed in an optimization task. We have approached this challenging resolution via two local search techniques, HC and TS, and a population-based technique, GA, which will be described below. To this end, let us first define the representation used, as well as possible neighborhood structures.

20.3.1 Representation and Evaluation

The matrix model provides a very natural way of representing the problem. Taking the model suggested by [9] each candidate solution is represented by a matrix $M = \{p_{ij}\}_{V \times T}$, where as mentioned before $p_{ij} \in \{1, \ldots, S\}$ indicates the number of copies of variation i in template T_j. We enforce the constraint $\sum_i p_{ij} = S$ for each template j. Notice the inherent symmetry of the representation: columns (i. e., templates) can be arbitrarily reordered. However, this representation breaks the symmetry inside the templates (i. e., it only matters how many but not which slots are used by each variation).

The initialization of the candidate solutions can be done randomly (i. e., assigning a random variation to each slot) or via some ad hoc heuristic. In the latter case, we can consider the process shown in Algorithm 20.1. It essentially amounts to a biased random initialization in which variation are given slots with a probability proportional to the fraction of the total demand they represent. A correction mechanism was also

Algorithm 20.1 Pseudo code of initialization under demand (ID).

```
p ← INITIALIZETEMPLATE(T, V)
    // Computes the fraction of the demand for each variation.
pq ← COMPUTEPROBABILITIES(Q)
for i ← 1 do T
    // Fills the i-th template according to pq.
    ROULETTEFILL(p, pq, i, V)
    // Computes a heuristic number of pressings for the i-th
    // template and updates probabilities according to the
    // remaining demand.
    R ← min_{j=1...V} {(1 − L_j)Q_j / p_ij | p_ij > 0}
    for j ← 1 do V
        Q_j ← Q_j − R * p_ij
    end for
    pq ← COMPUTEPROBABILITIES(Q)
end for
```

used to avoid that some variation be associated to no slot. This means enforcing the constraint

$$\forall i \in V : \sum_{j=1}^{T} p_{ij} > 0. \tag{20.3}$$

Given this representation, a simple neighborhood is considered by reassigning a single slot in a template to a different variation. Given the constraint mentioned previously, this means that a variation with nonzero appearances in a certain template must be selected (i. e., find a certain $p_{ij} > 0$), its count number has to be decreased, and subsequently the number of copies of another variation in the same template must be increased. For the sake of efficiency, the variation i whose number of copies is decreased is selected such that $\sum_j p_{ij} > 1$, since otherwise its total count number across all templates could drop to zero and hence no copy of it could be printed at all – recall Equation (20.3). Notice that the maximum size of the neighborhood is $\Theta(TV^2)$.

In order to evaluate a certain solution, the ILP model is solved by using `lp_solve` [2], a simplex-based software for linear and integer programming problems, and the returned value (the waste) is minimized. Notice that it may be possible that no feasible solution exists for an arbitrary configuration of templates, i. e., the target production cannot be reached within the desired tolerances. If the solver indicates that this is the case, the problem is relaxed by removing the overproduction constraint and a penalty term is added.

20.3.2 Metaheuristic Approaches

The hill-climbing (HC) approach and a tabu search [3,4] (TS) algorithm were defined on the basis of the neighborhood structure mentioned before.

The HC algorithms follow a steepest-descent procedure: the neighborhood of the current solution is partially explored, and the best solution is chosen unless it is worse than the current one; if this is the case, the current solution is a local optimum, and the process is restarted from a different point (randomly or heuristicly chosen) until the computational budget allocated is exhausted. Regarding the TS algorithm, the neighborhood is handled in the same way as in the algorithm HC, moving to the best nontabu neighbor even if it is worse than the current solution. A move is tabu if attempts to reverse a slot reassignment $\{p_{ij}^{++}, p_{kj}^{--}\}$ previously done and stored in the tabu list; here the notation p_{ij}^{++} (resp. p_{kj}^{--}) indicates that the number of copies of variation i (resp. k different from i) in template j have been increased (resp. decreased) in one occurrence. To prevent cycling, the tabu tenure – i. e., the maximum number of iterations a tabu move stays in the list – is chosen randomly in the range $[\beta/2, 3\beta/2]$, where $\beta = V^2 T$. The tabu status of a move can be overridden if the aspiration criteria is fulfilled, namely, finding a solution better than the current best solution found so far. After a number of n_ι evaluations (a parameter that we will set as a function of the total number of evaluations) with no improvement, the search is intensified, by returning to the best solution found so far.

In addition to the local search approaches previously mentioned, a genetic algorithm [1] (GA) is also considered. This GA handles a population of candidate solutions that are iteratively selected, bred and replaced. To be precise, selection is done by binary tournament, and replacement is done following a $(\mu, 1)$-policy, i. e., a new individual is generated and inserted in the population, replacing the worst one. As to breeding, it is done by recombination and mutation as usual. The recombination operator used is a variant of uniform crossover (UX) [8]: first, the templates are compared to find the best matching among them (recall the symmetry of the problem, and the fact that templates can be arbitrarily reordered). Subsequently, a template-level exchange of information is done by randomly choosing a template out of each pair of matched templates (one from each of the parents). Note that this amounts to a column-wise exchange in matrix $M = \{p_{ij}\}$. The mutation is handled in the same way as in the previously mentioned neighborhood.

A final remark must be made regarding the overall resolution approach: an incremental process is attempted by trying first to solve the problem with a low number (e. g., one) of templates, and then using the solutions found in this scenario as initial solutions (either for local search or for inclusion in the GA population) for the resolution of the problem with a higher number of templates. To this end, the candidate solutions must obviously be completed (e. g., a solution for the problem with $T = 1$ needs to be expanded with an additional template to be a tentative solution for the problem with $T = 2$). This is done following the same initialization techniques sketched above.

Table 20.1. Result for the TDP instances in [7].

	Sequential ILP solutions		
Problem	*No. of templates*	*Overall % dev.*	*Maximum % dev.*
cat food cartons	3	−2.59	−7.27
herbs cartons	3	−2.91	−8.70
magazine inserts	4	−2.49	−10.00

	Goal program adjustment of constraint programming solutions		
Problem	*No. of templates*	*Overall % dev.*	*Maximum % dev.*
cat food cartons	3	−1.4	−1.9
herbs cartons	2	−3.2	$+10/-6.7$
magazine inserts	3	−2.4	$+7.4/-8.3$

20.4 Experimental Results

Three instances of the problem, with the same demand requirements as those described in [7], have been considered for the experiments. Table 20.1 shows a summary of the results obtained by Proll and Smith (see [7, Tables 3 and 7 and Eqs. (7)–(10)]).

All of our algorithms have been run 20 times per problem instance. The number of evaluations for each scenario is subject to the number of variations and templates: $n_\nu = 1000 \cdot T \cdot V \cdot (V-1) \cdot \%_v$, where $\%_v$ represents the percentage of neighbors to be evaluated (note that a full exploration schema of the neighborhood is very costly) with $V = 50$ and $T = 4$, evaluating just 5 % of the neighborhood means to evaluate $4.9 \cdot 10^5$ neighbors – and thus a partial exploration policy was considered). For LS algorithms we set $n_\iota = n_\nu/10$. As to the GA, we set *population size* $= 100$, crossover and mutation probabilities $p_X = .9$, and $p_M = 1/(VT)$.

Tables 20.2 and 20.3 show the results obtained by applying HC and TS respectively. Initially the LS algorithms were executed with $\%_v = 5\,\%$ and $\%_v = 1\,\%$ respectively for the instances *herbs cartons* and *magazine insert*, but the performance was poor; as to the less complex scenario, that is to say *cat food*, we employed a full neighborhood exploration schema but again the results were really poor. After executing a number of preliminary experiments, we set $\%_v$ to 50 % for *cat food*, 10 % for *herbs cartons*, and 5 % for *magazine inserts*. Observe that TS provides the best results, and that both TS and HC provide better results in the instances *cat food* and *herbs cartons* but worse in the instance *magazine insert* (with 4 templates) than those given in [7]. The results for the GA are provided in Table 20.4. The GA matches the results obtained by the LS methods in the smaller instances but outperforms them in the most complex scenario (i. e., em Magazine Insert) producing here results close to those reported by Smith and Proll.

Table 20.2. Results for hill climbing (20 runs per instance).

Problem	*No. Template*	*Templates*	*Pressings*	*Overall % Dev.*	*Maximum % Dev.*
Cat Food	2	$T1_2^1$	157,143		
		$T1_2^2$	250,000	0.80	−3.85/1.79
	3	$T1_3^1$	7,143		
		$T1_3^2$	150,000		
		$T1_3^3$	250,000	0.14	−1.10/0.84
Herb. Cartons	2	$T2_2^1$	16,669		
		$T2_2^2$	65,554	2.98	−8.64/ 9.88
	3	$T2_3^1$	3,333		
		$T2_3^2$	16,666		
		$T2_3^3$	63,335	0.76	−4.76/ 7.41
Magaz. Inserts	3	$T3_3^1$	21,666		
		$T3_3^2$	135,000		
		$T3_3^3$	83,333	5.83	−10.00/ 66.67
	4	$T3_4^1$	12,500		
		$T3_4^2$	25,000		
		$T3_4^3$	125,000		
		$T3_4^4$	75,000	2.60	−4.41/ 50.00

$T1_2^1$=[0,0,0,0,0,2,7]
$T1_2^2$=[1,1,1,2,2,2,0]
$T1_3^1$=[0,1,1,0,0,7,0]
$T1_3^2$=[0,0,0,0,0,5,4]
$T1_3^3$=[1,1,1,2,2,0,2]
$T2_2^1$= [0,0,0,0,0,0,0,0,0,1,1,1,1,2,1,
1,1,1,1,2,2,2,2,5,2,2,2,2,5,5]
$T2_2^2$=[1,1,1,1,1,1,1,1,1,1,1,1,1,1,1,
1,1,1,1,1,1,1,1,1,3,3,3,3,3,3]
$T2_3^1$=[0,0,2,2,1,2,2,2,2,0,0,0,0,3,3,
3,0,3,3,1,1,1,1,2,1,1,1,1,2,2]
$T2_3^2$=[0,0,0,0,0,0,0,0,0,1,1,1,1,1,1,
1,2,1,1,2,2,2,2,1,6,6,2,2,1,5]
$T2_3^3$=[1,1,1,1,1,1,1,1,1,1,1,1,1,1,1,
1,1,1,1,1,1,1,1,2,2,2,3,3,4,3]
$T3$=There is no feasible solution

Table 20.3. Results for tabu search (20 runs per instance).

Problem	*No. Temp-late*	*Temp-lates*	*Pres-sings*	*Overall % Dev.*	*Maximum % Dev.*
Cat Food	2	$T1_2^1$	157,143		
		$T1_2^2$	250,000	0.80	−3.85/1.79
	3	$T1_3^1$	7,143		
		$T1_3^2$	150,000		
		$T1_3^3$	250,000	0.14	−1.10/0.84
Herb. Cartons	2	$T2_2^1$	16,400		
		$T2_2^2$	66,000	3.02	−8.44/10.00
	3	$T2_3^1$	3,274		
		$T2_3^2$	17,091		
		$T2_3^3$	63,090	0.73	−2.73/8.08
Magaz. Inserts	3	$T3_3^1$	38,250		
		$T3_3^2$	150,000		
		$T3_3^3$	50,000	4.00	−10.00/ 42.86
	4	$T3_4^1$	27,750		
		$T3_4^2$	38,250		
		$T3_4^3$	111,750		
		$T3_4^4$	55,000	2.59	−8.33/ 10.00

$T1_2^1$=$T1_3^1$=[0,0,0,0,0,2,7]
$T1_2^2$=$T1_3^2$=[1,1,1,2,2,2,0]
$T1_3^3$=[0,1,1,0,0,0,7]

$T2_2^1$= [0,0,0,0,0,0,0,0,0,1,1,1,1,1,1
1,2,1,1,2,2,2,2,1,6,2,2,2,5,5]

$T2_2^2$=[1,1,1,1,1,1,1,1,1,1,1,1,1,1,1
1,1,1,1,1,1,1,1,2,2,3,3,3,3,3]

$T2_3^1$=[0,0,2,2,2,2,2,2,2,0,0,0,0,3,3,
3,0,3,3,0,1,0,1,2,2,2,2,2,1,0]

$T2_3^2$=[0,0,0,0,0,0,0,0,0,1,1,1,1,1,1,
1,2,1,1,2,2,2,2,1,2,2,2,2,5,9]

$T2_3^3$=[1,1,1,1,1,1,1,1,1,1,1,1,1,1,1,
1,1,1,1,1,1,1,1,2,3,3,3,3,3,2]

$T3_3^1$ to $T3_3^3$ =There is no feasible solution

$T3_4^1$=[0,0,0,0,1,1,0,0,0,1,1,1,1,1,0,0,0,0,0,0,0,1,1,1,
0,0,1,2,2,1,2,2,2,2,2,1,0,3,0,0,3,0,2,1,1,0,0,0,3]

$T3_4^2$=[0,0,0,0,0,0,1,1,0,1,0,0,0,0,1,1,1,1,1,1,1,1,1,1,
2,2,0,1,1,0,0,0,0,0,0,0,2,0,2,2,0,2,1,0,2,1,4,4,0]

$T3_4^3$=[0,0,0,0,0,0,0,0,1,0,1,1,1,1,1,1,1,1,1,1,1,1,1,1,
0,0,1,0,0,1,1,1,1,1,1,1,1,1,1,1,1,1,1,2,1,2,0,2,2]

$T3_4^4$=[1,1,1,1,1,1,1,1,0,1,0,0,0,0,0,0,0,0,0,0,0,0,0,0,
2,2,1,2,2,1,1,1,1,1,1,2,1,1,1,1,1,1,1,0,1,0,4,0,2]

Table 20.4. Results for genetic algorithms (20 runs per instance).

Problem	*No. Temp-late*	*Temp-lates*	*Pres-sings*	*Overall % Dev.*	*Maximum % Dev.*
	2	$T1_2^1$	250,000		
Cat		$T1_2^2$	157,143	0.80	−3.85/1.79
Food	3	$T1_3^1$	250,000		
		$T1_3^2$	150,000		
		$T1_3^3$	7,143	0.14	−1.10/0.84
	2	$T2_2^1$	66,000		
Herb.		$T2_2^1$	16,400	3.02	−8.44/10.00
Cartons	3	$T2_3^1$	62,944		
		$T2_3^2$	17,348		
		$T2_3^3$	3,236	0.76	−2.36/8.49
	3	$T3_3^1$	150,000		
		$T3_3^2$	54,500		
Magaz.		$T3_3^3$	30,500	3.65	−9.75/10.00
Inserts	4	$T3_4^1$	140,000		
		$T3_4^2$	55,000		
		$T3_4^3$	30,000		
		$T3_4^4$	10,000	1.41	−6.00/10.00

$T1_2^1$=$T1_3^1$=[1,1,1,2,2,2,0]
$T1_2^2$=$T1_3^2$=[0,0,0,0,0,2,7]
$T1_3^3$=[0,1,1,0,0,0,7]
$T2_2^1$=$T2_3^1$= [1,1,1,1,1,1,1,1,1,1,1,1,1,1,1,
1,1,1,1,1,1,1,1,1,2,2,3,3,3,3]
$T2_2^2$=$T2_3^2$= [0,0,0,0,0,0,0,0,0,1,1,1,1,1,1,
1,1,1,2,2,2,2,2,1,6,2,2,2,5,5]
$T2_3^3$= [0,0,0,0,0,0,0,0,0,1,1,1,1,1,1,
1,1,1,2,2,2,2,2,1,6,2,2,2,5,5]
$T3_3^1$=$T3_4^1$=
[0,0,0,0,0,0,0,0,0,1,1,1,1,1,1,1,1,1,1,1,1,1,1,
1,1,1,1,1,1,1,1,1,1,1,1,1,1,1,1,1,1,1,1,1,1,1,1,1]
$T3_3^2$=$T3_4^2$=
[1,1,1,1,1,1,2,0,1,2,0,0,0,0,0,0,0,0,0,0,0,0,0,
1,0,1,1,0,1,1,1,1,1,1,1,2,1,1,1,1,1,0,1,1,2,2,3,3]
$T3_3^3$=$T3_4^3$=
[0,0,0,0,1,1,0,3,2,2,0,0,0,0,0,0,0,0,0,0,1,1,1,
0,1,0,0,1,0,0,1,1,1,1,1,0,1,1,2,2,1,3,2,2,0,4,2,3]
$T3_4^4$=
[0,0,0,1,0,0,0,1,0,0,0,0,0,0,1,1,1,1,1,1,1,0,0,0,
0,3,0,0,3,2,2,0,0,1,0,3,0,3,1,0,0,2,2,1,2,2,1,1,1]

20.5 Conclusions and Future Work

The application of metaheuristics to TDP has shown very encouraging and positive results. An empirical comparison of three different techniques (i. e., a hill climbing method, a tabu search algorithm, and a genetic algorithm) has been conducted in an incremental scenario. The results of the population-based approach are better than those of the LS approaches, and get very close to (and actually in two out of three instances improve) the results reported in the literature for this problem, without requiring manual intervention, as in [7].

We believe that further improved results can be attained with a hybrid proposal combining different algorithmic approaches, i. e., a memetic combination of HC/TS and GAs [5]. The form of this combination is an issue for further work.

Acknowledgments. This work is supported by project NEMESIS (TIN-2008-05941) of the Spanish Ministerio de Ciencia e Innovación, and project TIC-6083 of Junta de Andalucía.

References

[1] T. Bäck, D. Fogel, and Z. Michalewicz, *Evolutionary Computation 1: Basic Algorithms and Operators*, Institute of Physics Publishing, Bristol, 2000.

[2] M. Berkelaar, K. Eikland, and P. Notebaert, `lp_solve`, Open source (Mixed-Integer) Linear Programming system, 2004.

[3] F. Glover, Tabu Search – Part I, *ORSA Journal of Computing* **1** (1989), 190–206.

[4] F. Glover, Tabu Search – Part II, *ORSA Journal of Computing* **2** (1989), 4–31.

[5] P. Moscato and C. Cotta, A Modern Introduction to Memetic Algorithms, in: M. Gendrau and J.-Y. Potvin (eds.), *Handbook of Metaheuristics*, 2nd ed,, pp. 141–183, International Series in Operations Research and Management Science 146, Springer, New York, 2010.

[6] S. D. Prestwich, S. A. Tarim, and B. Hnich, Template Design under Demand Uncertainty by Integer Linear Local Search, *International Journal of Production Research* **44** (2006), 4915–4928.

[7] L. Proll and B. Smith, *ILP and Constraint Programming Approaches to a Template Design Problem*, School of Computer Studies, Report no. 97.16, 1998.

[8] G. Syswerda, Uniform Crossover in Genetic Algorithms, in: J. Schaffer (ed.), *3rd International Conference on Genetic Algorithms*, pp. 2–9, Morgan Kaufmann, San Mateo, CA, 1989.

[9] P. F. Uppsala, P. Flener, A. M. Frisch, B. Hnich, Z. Kızıltan, I. Miguel, and T. Walsh, Matrix Modelling, in: B. Smith, K. Brown, I. Philip Gent, and P. Prosser (eds.), *Workshop on Modelling and Problem Formulation. Conference on the Principles and Practice of Constraint Programming*, 2002.

Author information

David Rodríguez Rueda, Laboratorio de Computación de Alto Rendimiento (LCAR), Universidad Nacional Experimental del Táchira (UNET), San Cristóbal, Venezuela.
Email: `drodri@unet.edu.ve`

Carlos Cotta, ETSI Informática, Universidad de Málaga, Campus de Teatinos, Málaga, Spain.
Email: `ccottap@lcc.uma.es`

Antonio J. Fernández-Leiva, ETSI Informática, Universidad de Málaga, Campus de Teatinos, Málaga, Spain.
Email: `afdez@lcc.uma.es`

Chapter 21

A Comparison of Simulated Annealing and Genetic Algorithm Approaches for Cultivation Model Identification

Olympia Roeva

Abstract. Simulated annealing (SA) and genetic algorithms (GA) are two stochastic methods that are currently in wide use for solving difficult optimization problems. These algorithms are often viewed as quite distinct, competing paradigms in the field of modern heuristics. This chapter compares the performance of GA and SA for parameter identification of a nonlinear mathematical model of an E. coli cultivation process. Parameter identification is carried out using a real experimental data set from the cultivation process. The results show that GA produces solutions that are identical to or better than those obtained with SA, but that SA achieves its solutions much more quickly.

Keywords. Simulated Annealing, Genetic Algorithm, Cultivation process, Identification.

Mathematics Subject Classification 2010. 37N25, 90C31, 97M60.

21.1 Introduction

Due to the complexity of many optimization problems, exact algorithms often perform very poorly. As a result, heuristic, i. e., metaheuristic, algorithms are conspicuously preferable in practical applications [3, 9, 10]. Simulated annealing (SA) and genetic algorithms (GA) are often viewed as quite separate, competing paradigms in the field of modern heuristics. GAs are probabilistic search algorithms, which start with an initial population of likely problem solutions and then evolve towards better solutions [4]. They are based on the mechanics of natural genetics and natural selection. SA is a stochastic relaxation technique [5]. SA is so named by analogy to the annealing of solids, in which a crystalline solid is heated to its melting point and then allowed to cool gradually until it is again in the solid phase at some nominal temperature. At the absolute zero final temperature, the resulting solid achieves its most regular crystal configuration, corresponding to a (global) minimal value of the system's energy.

Theoretically, SA and GA are quite closel relatived, and much of difference between them is superficial. The two approaches are usually formulated in ways that look very different and use very different terminology. Basically, SA can be thought of as a GA where the population size is only one. The current solution is the only individual in the population. Since there is only one individual, there is no crossover but only mutation. This is in fact the key difference between SA and GA. While SA creates a new solution by modifying only one solution with a local move, GA also creates solutions by combining two different solutions. Whether this actually makes the algorithm better or worse, is not straightforward, but depends on the problem and the representation.

In [8] it was observed that SA is more robust and reliable than GA when used for optimization of low-temperature gas separation processes. In [2], for the clustering rectangles problem the SA and TS algorithms produce similar solutions and both outperform GA. In [7] a comparison of computation time shows that SA has the minimum computation time and GA has the maximum one. In this chapter, for the first time both GA and SA are applied and compared for a parameter identification of a nonlinear mathematical model of an E. coli MC4110 fed-batch cultivation process.

The chapter is organized as follows. A brief description of the GA and SA algorithms applied here is proposed in Sections 21.2 and 21.3. In Section 21.4 a dynamic nonlinear model of E. coli fed-batch cultivation process is described. The obtained numerical results and discussion are presented in Section 21.5. Concluding remarks are made in Section 21.6.

21.2 Genetic Algorithm

A pseudo code of a GA is presented as

---

begin
    $i = 0$
    Initial population $P(0)$
    Evaluate $P(0)$
    while (not done) **do** (test for termination criterion)
    begin
        $i = i + 1$
        Select $P(i)$ from $P(i-1)$
        Recombine $P(i)$
        Mutate $P(i)$
        Evaluate $P(i)$
    end
end

The population at time t is represented by the time-dependent variable P, with the initial population of random estimates being $P(0)$. Here, each decision variable in the parameter set is encoded as a binary string (with precision of binary representation = 20). The initial population is generated using a random number generator that uniformly distributes numbers in the desired range. The objective function (see Equation (21.8)) is used to provide a measure of how individuals have performed in the problem domain. Another function, the fitness function, is normally used to transform the objective function value into a measure of relative fitness. A commonly used transformation is that of proportional fitness assignment:

$$F(x_i) = \frac{f(x_i)}{\sum_{i=1}^{N_{ind}} f(x_i)} \tag{21.1}$$

where f is the objective function, F is the resulting relative fitness, N_{ind} is the population size and x_i is the phenotypic value of individual i.

To solve the considered optimization problem the population size is chosen to be 100 after several algorithm performance pretests. In the same manner the number of generations is set at 200. The GA presented here employs a "roulette wheel" mechanism to probabilistically select individuals based on some measures of their performance. Here a double point crossover with crossover probability of 0.7 is accepted. Mutation is randomly applied with low probability, in the range 0.01. A generation gap of 0.97 is chosen, and fitness-based reinsertion is used. Since GA is a stochastic search method, it is difficult to formally specify the convergence criteria. As the fitness of a population may remain static for a number of generations before a superior individual is found, the application of conventional termination criteria becomes problematic. A common practice is to terminate the GA after a prespecified number of generations.

21.3 Simulated Annealing

A pseudo code of a SA is presented as

Find initial solution (by generating it randomly)
Set initial value for the control parameter $T = T_0$
Set a value for r, the rate of cooling parameter
    $j = 0$
    Generate (at random) a new solution S'
    Calculate the difference in cost: $\Delta = \text{cost}(S') - \text{cost}(S)$
    Examine the new solution and decide: accept or reject
    If accepted, it becomes the current solution;
    otherwise, keep the old one; $j = j + 1$
    Reduce the temperature and generate a new solution
Until some stopping criterion applies

The algorithm generates a random trial point. Initial temperature is 100. The SA chooses the distance of the trial point from the current point by a probability distribution with a scale depending on the current temperature. The trial point distance distribution could be set as function that generates a point based on the current point and the current temperature using different distributions. Here Boltzman distribution is used [5]. The algorithm determines whether the new point is better or worse than the current point. The algorithm accepts a worse point based on an acceptance function. The probability of acceptance is

$$\frac{1}{1+\exp\left(\frac{\Delta}{\max(T)}\right)}, \tag{21.2}$$

where Δ = new objective – old objective, T = current temperature. The algorithm systematically lowers the temperature, storing the best point found so far. The function that the algorithm uses to update the temperature is: $T = T_0 * 0.95^r$, where r denotes the annealing parameter. Reannealing sets the annealing parameters to values lower than the iteration number, thus raising the temperature in each dimension. The annealing parameters depend on the values of estimated gradients of the objective function in each dimension:

$$r_i = \log\left(\frac{T_0}{T_i}\frac{\max\limits_j(s_j)}{s_i}\right), \tag{21.3}$$

where: r_i = annealing parameter for component i, T_0 = the initial temperature of component i, T_i = the current temperature of component i, and s_i = the gradient of objective in direction i × the difference of bounds in direction i.

In the considered SA algorithm the number of points accepted before reannealing is 100. The algorithm stops when the average change in the objective function is small relative to the predefined tolerance.

21.4 E. coli MC4110 Fed-Batch Cultivation Process Model

The mathematical model of the considered cultivation process can be represented by the following dynamic mass balance equations [6]:

$$\frac{dX}{dt} = \mu_{\max}\frac{S}{k_S+S}X - \frac{F_{in}}{V}X \tag{21.4}$$

$$\frac{dS}{dt} = -\frac{1}{Y_{S/X}}\mu_{\max}\frac{S}{k_S+S}X + \frac{F_{in}}{V}(S_{in}-S) \tag{21.5}$$

$$\frac{dpO_2}{dt} = -\frac{1}{Y_{pO_2/X}}\mu_{\max}\frac{S}{k_S+S}X + k_La(pO_2^* - pO_2) - \frac{F_{in}}{V}pO_2 \tag{21.6}$$

$$\frac{dV}{dt} = F_{in} \tag{21.7}$$

where X = the biomass concentration, [g/l],S = the substrate concentration, [g/l], pO_2 = dissolved oxygen concentration, [%], pO_2^* = saturation concentration of dissolved oxygen, [%] F_{in} = feeding rate, [l/h] V = bioreactor volume, [l], S_{in} =

substrate concentration in the feeding solution, [g/l], $\mu_{\max}$ = maximum value of the specific growth rate, $[h^{-1}]$, k_S = saturation constant, [g/l], $k_L a$ = volumetric oxygen transfer coefficient, $[h^{-1}]$, and $Y_{S/X}$ and $Y_{pO_2/X}$ = yield coefficients, [-].

21.5 Numerical Results and Discussion

For comparison of the performance of the GA and SA algorithms presented here, a parameter identification of model parameters ($\mu_{\max}$, k_S, $Y_{S/X}$, $Y_{pO_2/X}$, $k_L a$, pO_2^*) is performed. The parameter identification is done using real experimental data from the E. coli fed-batch cultivation process [1]. Based on [6], a two stage parameter identification procedure is used. First, parameters $\mu_{\max}$, k_S and $Y_{S/X}$ are estimated, using Equations (21.4), (21.5), and (21.7), and second, parameters $Y_{pO_2/X}$, $k_L a$ and pO_2^* are evaluated, using Equations (21.4)–(21.7) and already estimated on the first step model parameters. The objective function is presented as a minimization of a distance measure J between experimental and model predicted values of state variables, represented by the vector $\mathbf{y}$:

$$J = \sum_{i=1}^{n} \sum_{j=1}^{m} \{[\mathbf{y}_{\exp}(i) - \mathbf{y}_{\mathrm{mod}}(i)]_j\}^2 \to \min, \tag{21.8}$$

where n = the number of data for each state variable, m = the number of state variables, $\mathbf{y}_{\exp}$ = the known experimental data, and $\mathbf{y}_{\mathrm{mod}}$ = model predictions with a given set of the parameters.

Parameters of the two algorithms GA and SA were tuned based on several pre-tests according to the optimization problem considered here. After tuning procedures, the main algorithm parameters are set to the optimal settings (Section 21.2). Computer specification to run all identification procedures are Intel Core 2 Qard CPU Q8200 @ 2.33 GHz, 3.5 GB Memory, Windows XP operating system and Matlab 7.5 environment. Because of the stochastic characteristics of the applied algorithms, a series of 30 runs for each algorithm is performed. For comparison of the GA and SA, the best, the worst, and the average results of the 30 runs for the J value and execution time are watched. On the first step of the identification procedure the following parameters values are estimated:

$$\begin{aligned} &\text{GA:} \quad \mu_{\max} = 0.4796, \quad k_S = 0.0162, \quad Y_{S/X} = 2.0137,\\ &\text{SA:} \quad \mu_{\max} = 0.4864, \quad k_S = 0.0150, \quad Y_{S/X} = 1.9088. \end{aligned}$$

These are the best estimations and were used to perform the second parameter identification step. This is more complicated step. A few mathematical models of E. coli fed-batch cultivation process considered dissolved oxygen dynamics. Thus, the obtained results are more interesting. The obtained numerical results are presented in Table 21.1. In the table are listed the average and the best results. Some graphical results about algorithms performance are presented in Figures 21.1 and 21.2.

Table 21.1. Results from parameter identification – second step.

Parameters	*Average GA*	*Best GA*	*Average SA*	*Best SA*
execution time, s	213.5737	195.1406	169.2982	88.9688
J value	0.1367	0.11025	0.1494	0.11041
$1/Y_{pO_2/X}$	7.8215	7.3271	8.1277	7.1884
k_La	21.1521	20.8503	23.6376	20.5477
pO_2^*	21.2833	21.2879	21.2800	21.2863

The obtained results show that the algorithms produce the same estimations with more than 85 % coincidence. The execution time is a key element in empirical comparisons of the metaheuristic algorithms considered here. The obtained results show clearly that the execution time of the GA is longer than that of the SA. The best result from the SA is achieved for 88.9688 s. Compared to GA performance (195.1406 s), SA is two times faster. Average results for the execution time is identical: 169.2982 s for SA and 213.5737 s for GA. For SA, execution time only depends on the problem size and the number of iterations. For GA the execution time does not merely depend on the problem size and the number of iterations, but also depends on the population size and representation. In the same time for the achieved values of the objective function considered, no significant difference is found between the GA and SA. In the case of the best results, almost identical values of J are obtained: 0.11025 for GA vs. 0.11041 for SA (see Table 21.1 and Figures 21.1 and 21.2). In Figure 21.3 the comparison between simulated and real experimental data for the process variable (dissolved oxygen) is presented. The graphical results show very good correlation between the experimental and model predicted data for both metaheuristic algorithms.

The performance of both algorithms is mainly impacted by the parameter set selection. The GA demanded more carefully tuning (choosing the population size, the crossover and mutation probabilities, and generation gap) than SA which has been applied with standard options only. From the practical viewpoint, it should be noted that for some problems, evaluating solutions that are near an existing solution may be very efficient, which may give a large advantage performance to SA, when compared to GA, if evaluating recombined solutions are not so efficient.

21.6 Conclusion

In this Chapter two metaheuristics, GA and SA, for parameter identification of nonlineal model of cultivation process are compared. A real E. coli MC4110 fed-batch cultivation process is considered. The mathematical model is presented by a system of ordinary differential equations, describing the regarded cultivation process variables. A particular procedure for model parameter identification is applied using GA and SA. Numerical and simulation results reveal that correct and consistent results can

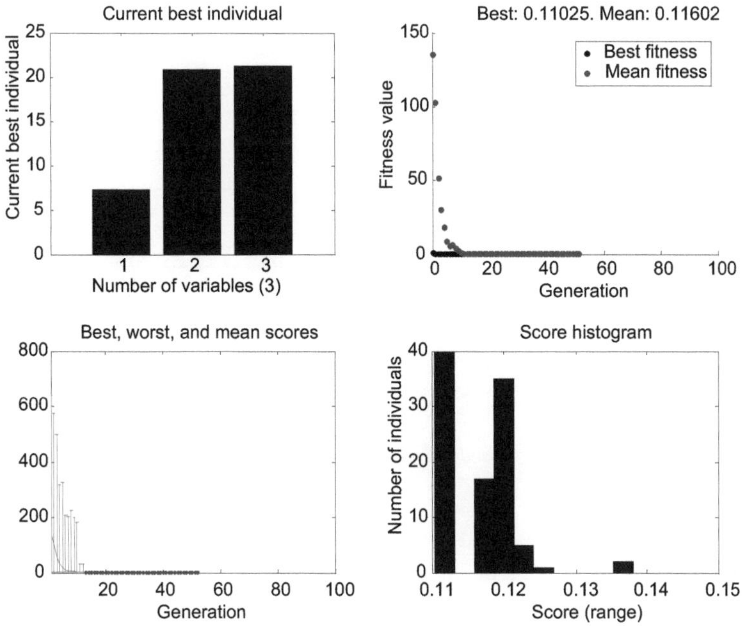

Figure 21.1. Best GA result.

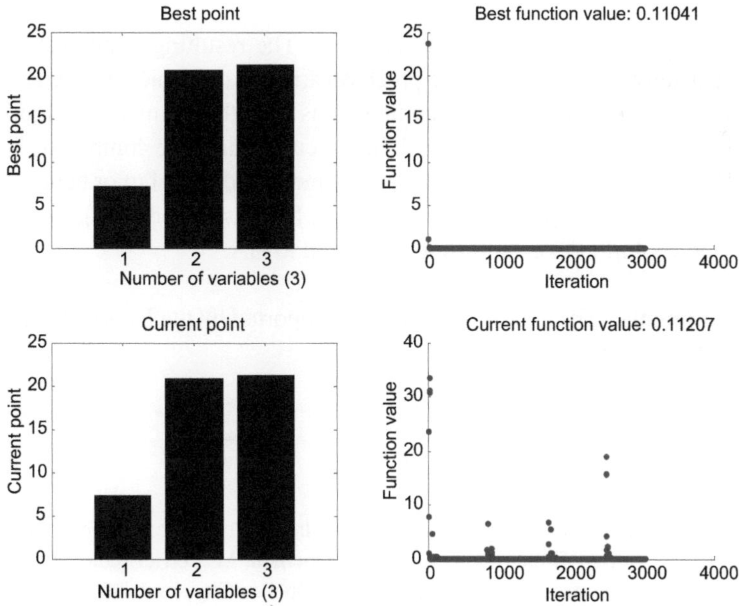

Figure 21.2. Best SA result.

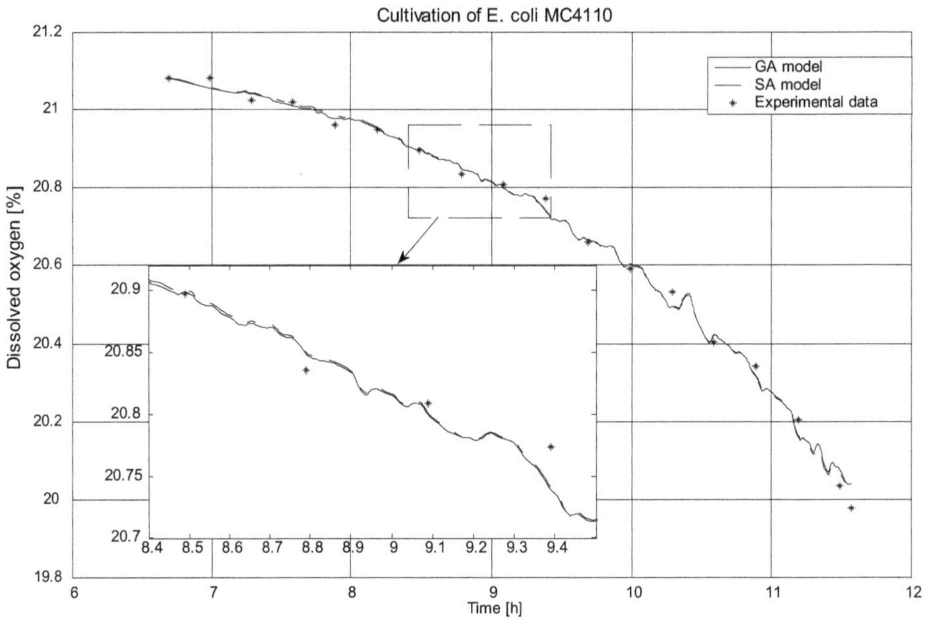

Figure 21.3. Comparison between model predicted and real experimental data.

be obtained using both metaheuristic searchers. The resulting mathematical models predict adequately and accurately the process variables dynamics during E. coli cultivation. Effectiveness and efficiency comparisons of both optimization algorithms are presented. The algorithms accuracy and the execution time are compared. The results show that GA obtains values of the objective function identical to or better than those obtained by SA. Regarding the execution time, SA has the advantage over GA. SA obtains a good solution in a short time compared to GA.

Acknowledgments. The author's research is supported by the National Science Fund Grants No. DMU 02/4 and DID 02-29.

References

[1] M. Arndt and B. Hitzmann, Feed Forward/Feedback Control of Glucose Concentration during Cultivation of Escherichia coli, in: D. Dochain, M. Perrier (eds.), *Proc. 8th IFAC Int. Conf. on Comp. Appl. in Biotechn. Canada*, Quebec City, June 24–27, 2001, pp. 425–429, 2001.

[2] E. Burke and G. Kendall, Comparison of Meta-Heuristic Algorithms for Clustering Rectangles, *Computers and Industrial Engineering* **37**(1–2) (1999), 383–386.

[3] S. Fidanova, E. Alba, and G. Molina, Hybrid ACO Algorithm for the GPS Surveying Problem, in: I. Lirkov, S. Margenov, and J. Wasniewski (eds.), *Proc. 7th International Conference, LSSC 2009*, Sozopol, Bulgaria, June 2009. revised papers, Large Scale Scientfic Computing 5910, pp. 318–325, Springer, Berlin, 2010.

[4] D. Goldberg, *Genetic Algorithms in Search, Optimization and Machine Learning*, Addison-Weslcy, Reading, MA, 1989.

[5] S. Kirkpatrick, C. D. Gelatt, and M. P. Vecchi, Optimization by Simulated Annealing, *Science (New Series)* **220**(4598) (1983), 671–680.

[6] O. Roeva, Parameter Estimation of a Monod-type Model based on Genetic Algorithms and Sensitivity Analysis, in: I. Lirkov, S. Margenov, and J. Wasniewski (eds.), *6th International Conference, LSSC 2007*, Sozopol, Bulgaria, June 2007, revised papers, pp. 601–608, Lecture Notes in Computer Science 4818. Springer, Berlin, 2008.

[7] W. P. Syam and I. M. Al-Harkan, Comparison of Three Meta Heuristics to Optimize Hybrid Flow Shop Scheduling Problem with Parallel Machines, *World Academy of Science, Engineering and Technology* **62** (2010), 271–278.

[8] N. Tahouni, R. Smith, and M. H. Panjeshahi, Comparison of Stochastic Methods with Respect to Performance and Reliability of Low-temperature Gas Separation Processes, *The Canadian Journal of Chemical Engineering* **88**(2) (2010), 256–267.

[9] R. Umarani and V. Selvi, Particle Swarm Optimization: Evolution, Overview and Applications, *Int. J. of Engineering Science and Technology* **2**(7) (2010), 2802–2806.

[10] M. K. Yusof and M. A. Stapa, Achieving of Tabu Search Algorithm for Scheduling Technique in Grid Computing using GridSim Simulation Tool: Multiple Jobs on Limited Resource, *Int. J. of Grid and Distributed Computing* **3**(4) (2010), 19–31.

Author information

Olympia Roeva, Institute of Biophysics and Biomedical Engineering, Bulgarian Academy of Sciences, Sofia, Bulgaria.
Email: `olympia@biomed.bas.bg`

Chapter 22

Monte Carlo Investigations of Electron Decoherence due to Phonons

Philipp Schwaha, Mihail Nedjalkov, Siegfried Selberherr, and Ivan Dimov

Abstract. We investigate alternative Monte Carlo algorithms for simulation of the decoherence of entangled electron states due to scattering by phonons. We begin with a weighted single particle approach, which requires an estimator encompassing all the time steps to be held in memory. A second algorithm is obtained by synchronously evolving an ensemble of all particles together. While these algorithms are identical from a theoretical point of view, their implementation and run-time behavior differ significantly. A third algorithm exploits the idea of indistinguishable particles which condenses the information connected to particle location by using phase space cells. An increased number of particles can be considered in this way without exceeding memory constraints. However, an additional source of error arises due to numerical scattering – a consequence of consecutive averaging in the phase space cells. An estimate for the viability of this procedure is provided by a comparison with results obtained from the previous algorithms. Furthermore, the influence of the size of the phase space cells is investigated.

Keywords. 68W20, 68W40.

Mathematics Subject Classification 2010. Monte Carlo Algorithms, Coherent and Decoherent Evolution, Quantum System.

22.1 Introduction

Quantum computers provide an alternative to the current computation devices based on classical charge transport. They rely on fundamental physical processes such as superposition, entanglement, uncertainty, and interference. The research on quantum computing is mainly concerned with the possible speed-up, quantum complexity bounds, and construction of optimal quantum algorithms [1]. The basic unit of quantum information is the qubit, upon which logical operations act, and which are provided by quantum gates. The representation of a qubit can make use of basis quantum states. For example, for a system comprised of two states $|0\rangle$ and $|1\rangle$ any normalized superposition

$$|\psi\rangle = \alpha|0\rangle + \beta|1\rangle \tag{22.1}$$

is a legitimate qubit. While at first glance this superposition seems to closely resemble the classical case, there is the key difference that whereas a classical system can be in exactly one particular state, a quantum computer consisting of n qubits can be in an arbitrary superposition of up to 2^n different states simultaneously. The conservation of the quantum nature of a state as described by Equation (22.1) is paramount for quantum computers. This restricts the describability of the evolution by unitary automorphisms. To clarify the difference we recall that expectation values $\langle A\rangle$ of physical quantities A presented by a Hermitian operator $\hat{A}$ are obtained by the trace operation:

$$\langle A\rangle = Tr(\hat{A}\hat{\rho}) = \sum_{i=0,1} \langle i|\hat{A}\hat{\rho}|i\rangle; \qquad Tr(\hat{\rho}) = 1. \tag{22.2}$$

The density operator $\hat{\rho}$ is defined with the help of (22.1):

$$\hat{\rho} = |\psi\rangle\langle\psi| = |\alpha|^2|0\rangle\langle 0| + |\beta|^2|1\rangle\langle 1| + \alpha\beta^*|0\rangle\langle 1| + \alpha^*\beta|1\rangle\langle 0| \tag{22.3}$$

The two states $|0\rangle$ and $|1\rangle$, can be discriminated from each other by a detector due to their orthogonality. The physical quantity "expectation value for the first basis state" in (22.1), for example, expressed by the operator $\hat{\rho}_0 = |0\rangle\langle 0|$, evaluates to $|\alpha|^2$ using the trace operation, while for the second state it yields $|\beta|^2$. The relation $|\alpha|^2 + |\beta|^2 = 1$ allows to interpret these values as probabilities, which implies the normalization of (22.1) and ensures the last equality in (22.2). Thus, the first two terms represent a classical state comprised by expectation probabilities, while the rest of the terms in (22.3) reflect the coherence of the state by enabling interference effects.

A deviation from a coherent evolution due to interactions with the environment may turn an initially quantum system into a classical one, a process known as decoherence. As the latter destroys the coherence terms, it is the biggest issue for the realization and use of quantum computers and quantum devices in general. Many of the features of classicality are actually induced in quantum systems by their environment [2]. The evolution of an initial electron wave packet subject to scattering does not follow the natural process of spreading, as in the case under purely coherent evolution; the corresponding density matrix shrinks around the line $x = x'$, revealing a classical localization [3]. Decoherence of semiconductor electrons due to scattering by phonons and impurities was first demonstrated by Monte Carlo simulations of the evolution of a single wave packet [4]. Recently the decoherent evolution of an entangled, according to (22.1), state, comprised by two Gaussian wave packages $e^{-(x\pm a)^2/2\sigma^2}e^{ibx}$ has been studied in terms of a Wigner distribution f_w [5]. The corresponding initial function

$$f_w^0(x,k_x) = Ne^{-(k_x-b)^2\sigma^2}\left(e^{-\frac{(x-a)^2}{\sigma^2}} + e^{-\frac{(x+a)^2}{\sigma^2}} + e^{-\frac{x^2}{\sigma^2}}cos\left((k_x-b)2a\right)\right) \tag{22.4}$$

given by two packets and an oscillatory term is shown in Figure 22.1. Equilibrium is assumed in the other two directions of the wave vector space, so that

$$\frac{\hbar^2}{2\pi mkT}e^{-\frac{\hbar^2(k'^2_y+k'^2_z)}{2mkT}}$$

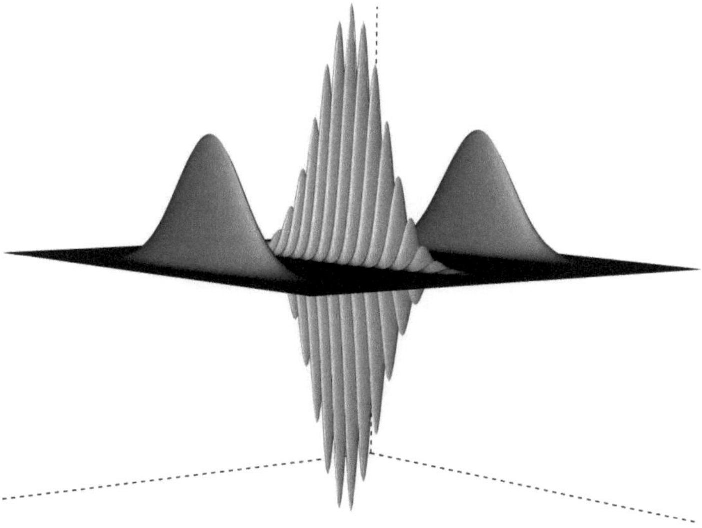

Figure 22.1. Entangled wave packets used as initial condition.

multiplies (22.4) to give the initial condition $f_w^0(x, \mathbf{k})$ in the free evolution equation

$$\left(\frac{\partial}{\partial t} + \frac{\hbar k_x}{m}\frac{\partial}{\partial x}\right) f_w(x, \mathbf{k}, t) = \int d\mathbf{k}' f_w(x, \mathbf{k}', t) S(\mathbf{k}', \mathbf{k}) - f_w(x, \mathbf{k}, t)\lambda(\mathbf{k}) \quad (22.5)$$

Phase-breaking processes are introduced by the Boltzmann scattering operator $S(\mathbf{k},\mathbf{k}')$, the rate for a transition from $\mathbf{k}$ to $\mathbf{k}'$. $\lambda(\mathbf{k}) = \int d\mathbf{k}' S(\mathbf{k}.\mathbf{k}')$ is the total out-scattering rate. The choice of $2\sigma^2 = \hbar^2/(2mkT)$ and $b = 0$ gives rise to a Maxwell–Boltzmann equilibrium distribution, which minimizes the effect of the phonons on the change in the shape of the wave vector distribution. Figure 22.2 shows, how scattering effectively destroys the oscillatory term, causing a collapse of f_w towards equilibrium.

These investigations pose requirements for relevant particle algorithms, capable of resolving the fine structure of f_w. We present three Monte Carlo algorithms for solving (22.5) and compare their numerical peculiarities.

22.2 The Algorithms

The comparison conveniently uses the fact that without scattering the initial momentum distribution defined by

$$f(k_x) = \int dx dk_y dk_z f_w(x, k_x, k_y, k_z), \quad (22.6)$$

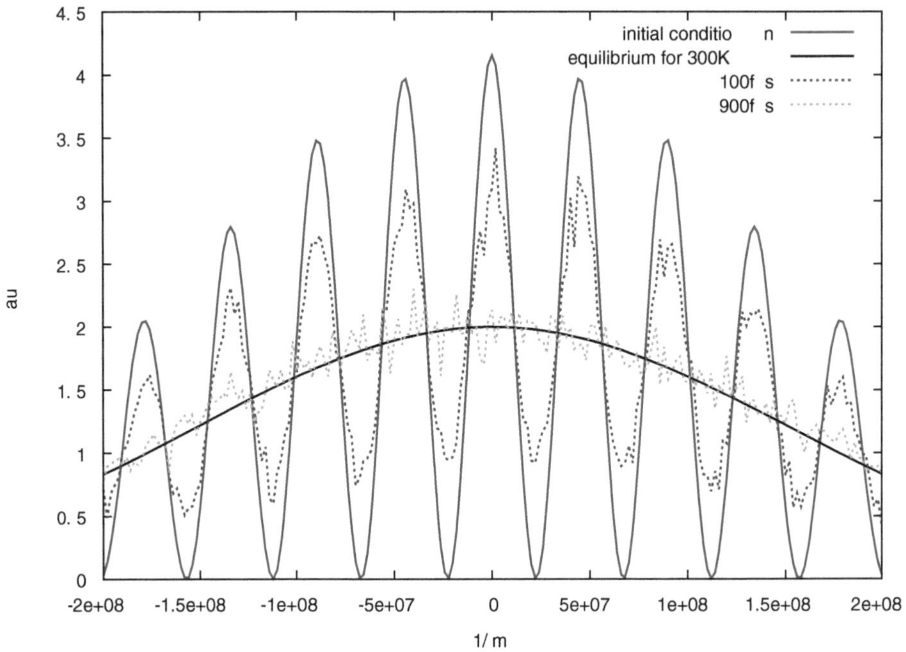

Figure 22.2. The momentum distribution, $f(k_x)$, (22.6), decays due to the scattering by phonons from the initial condition to the thermal equilibrium in approximately 1ps, $T = 300K$.

remains frozen in the wave vector space. The algorithms utilize estimators defined on a grid in the phase space. These estimators are updated by the particles in their vicinity. Each of the particles is initialized at a phase space position corresponding to the spatial coordinate and momentum provided by the initial condition f_{w}. The particle's weight is set according to the value of f_{w} in the initialization point and can thus be positive or negative. The particles are then evolved along Newton trajectories according to the flow defined by the system's Hamiltonian.

As time is evolved by time step increments, the weight is recorded at the estimators closest to the position to which the particle has evolved.

22.2.1 Algorithm A

Algorithm A utilizes the same grid structure as used for the estimators to choose the initial phase space position of the generated particles. The state of the particle is comprised by the position and momentum coordinates and its weight. Each particle is then evolved through all of the time steps individually and the weights are recorded. This requires that the estimators for all time steps must be held in memory during the whole

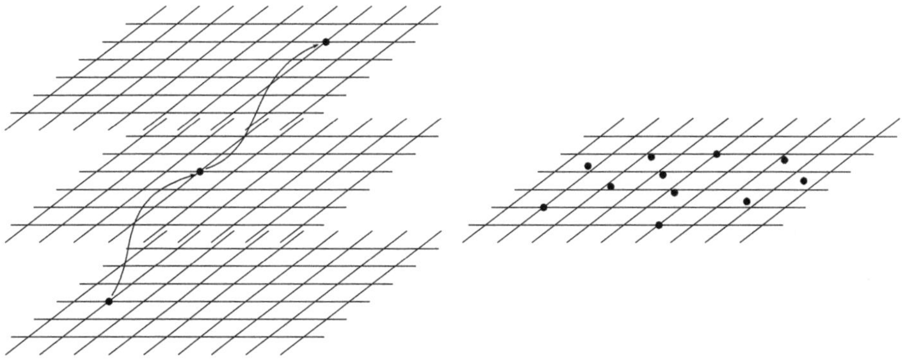

Figure 22.3. Schematic illustrations of Algorithms A (left) and B (right). Algorithm A needs to maintain the estimators for all time steps within memory, whereas Algorithm B must store the ensemble of particles and a single estimator for the current time step.

simulation. The amount of required memory can be estimated by

$$\text{sizeof}\,(\textit{estimator grid}) \times (\#\ \textit{time steps}) + \text{sizeof}\,(\textit{particle state})$$

22.2.2 Algorithm B

Algorithm B utilizes the same grid structure for the spawning of particles as Algorithm A, but instead of evolving the particles individually, they are evolved as an ensemble. Technically, this only requires the interchange of the order in which time and particles are iterated. In contrast to Algorithm A, only the estimator for a single time step must be held in memory. This, however, comes at the price that the entire ensemble of particles needs to be held. The amount of required memory can be estimated by

$$\text{sizeof}\,(\textit{particle state}) \times (\#\ \textit{particles}) + \text{sizeof}\,(\textit{estimator grid})$$

22.2.3 Algorithm C

In contrast to Algorithms A and B, Algorithm C abandons the use of identical grids for the creation of particles as well as estimators. The algorithm exploits the idea of indistinguishable particles within a grid cell. The particle state after any time step is now presented by the number associated to the grid cell, and the weight. At the beginning of a time step the particle starting position is chosen using a uniform random distribution within the cell. The evolution is then performed as in Algorithm B, as well as the recording of estimates on a grid. After the evolution, at the end of the time step, the particle is absorbed by the end grid cell, where the weights of the particles are recorded. The procedure of randomization of the initial coordinates is addressed as

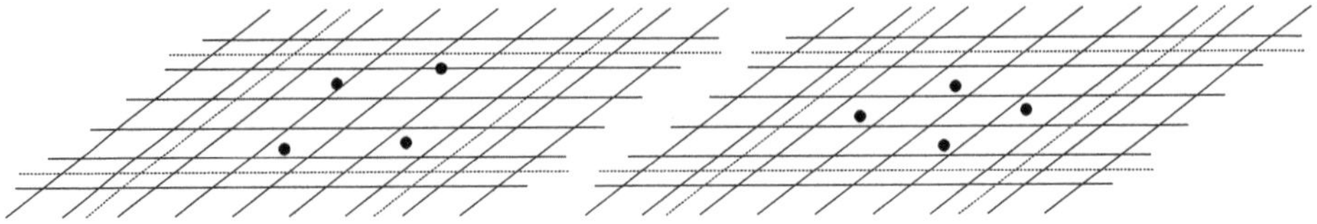

Figure 22.4. Algorithm C uses distinct cells (delimited by dotted lines) to hold the particles. The size of these cells is independent of the estimator grid. Two different realizations of a cell containing four particles are shown.

numerical scattering. In contrast to phonon scattering it is nonlocal in position, and no conservation rules are applied. The effect of the numerical scattering must disappear in the liming case of vanishing cell volumes, since then Algorithm C tends towards the procedure of Algorithm B, as the randomization is suppressed and individual positions are tracked once more. The advantage of Algorithm C is that particle position and momenta are not stored, as replaced by the number of the cell. This algorithm is very convenient for Wigner particle generation-annihilation schemes [6], where huge numbers of particles may appear in the simulations. Moreover, the algorithm can be generalized to store the total weight in the cell, which will be discussed elsewhere. In the following we investigate the effect of the numerical scattering as a grid size depending source of decoherence. The amount of required memory can be estimated by

$$(\#\, cells) \times \text{sizeof}\,(weight) + \text{sizeof}\,(particle\ state) + \text{sizeof}\,(estimator\ grid)$$

Figure 22.4 shows the cell structure as well as the estimator grid. Each cell records the individual weights of the particles within, while the positions within the cell are chosen at random. Thus the two configurations shown in Figure 22.4 are both identical with regard to the cells. Since the estimator grid can be adjusted independently of the cell structure, the estimators distinguish between the two configurations.

Simulations are presented, where Algorithms A and Algorithms B result in a perfect match and are used as reference A/B. The grid consists of 4000 points for the x axis and 3000 points for the k_x axis. Thus, using the previously given formulae for 100 time steps and a particle state requiring 32 Bytes, the memory requirements for Algorithm A can be determined to be 8.94 GiB, while Algorithm B requires 139.14 MiB. The finest cell structure, comprised of 12×10^6 cells, used in deployments of Algorithm C, consumes 183.11 MiB of memory, while the coarsest, corresponding to 12×10^4 cells, only requires 92.47 MiB. It can be seen that for 100 time steps Algorithm A requires excessive amounts of memory, when compared to the other algorithms, especially as Algorithm B produces matching results. The variation of the cell sizes allows an adjustment of the memory requirements of Algorithm C.

Figure 22.6 demonstrates the effect the cell size has on the reconstruction of the densities after a 10 fs time step. The main peaks are well resolved without regard for

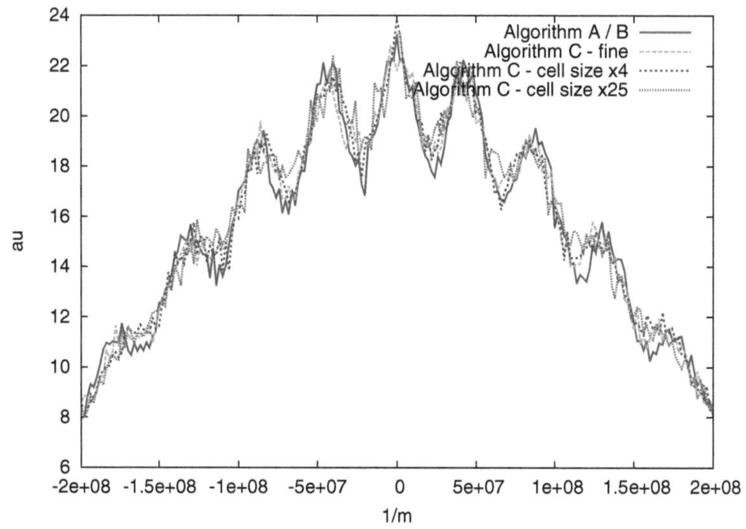

Figure 22.5. With phonon scattering Algorithm C manages to reproduce the results of Algorithms A/B as seen here after 300fs. This indicates that the effects of numerical scattering due to cell size are suppressed by the physical scattering due to phonons.

the cell's sizes. However, the quantum region in between the center of mass show spurious fluctuations in the densities, which are nonnegative quantities.

Figure 22.7 shows the distribution of the wave vectors after 700 fs of coherent evolution, including results for different sizes of the cells used in Algorithm C. For small cell sizes, Algorithm C correctly reproduces the frozen wave vector distribution of Algorithm A and Algorithm B. As the cell size increases by a factor of four, a spurious spike centred at 0 is encountered, while the remainder still adequately follows the reference results. A volume 25 times larger than the initial cells ensures a well-pronounced effect of decoherence due to the numerical scattering, similar to that caused by phonons in Figure 22.2.

Inclusion of scattering due to phonons in the algorithms leads to the results shown in Figure 22.5. Again, all algorithms are generally capable of reproducing the reference results. Phonon scattering dominates the evolution, overriding the effect of the numerical counterpart. This is important, since it allows for larger cell tolerance in mixed mode quantum simulations.

We have presented different Monte Carlo algorithm to compute the results of coherent and decoherent evolution of a quantum system. It has been found that a cell-based algorithm is suitable for these calculations, provided that the cells are chosen sufficiently fine in the coherent case. In mixed mode transport physical scattering dominates numerical effects, which allows a tolerance in the choice of cell volumes.

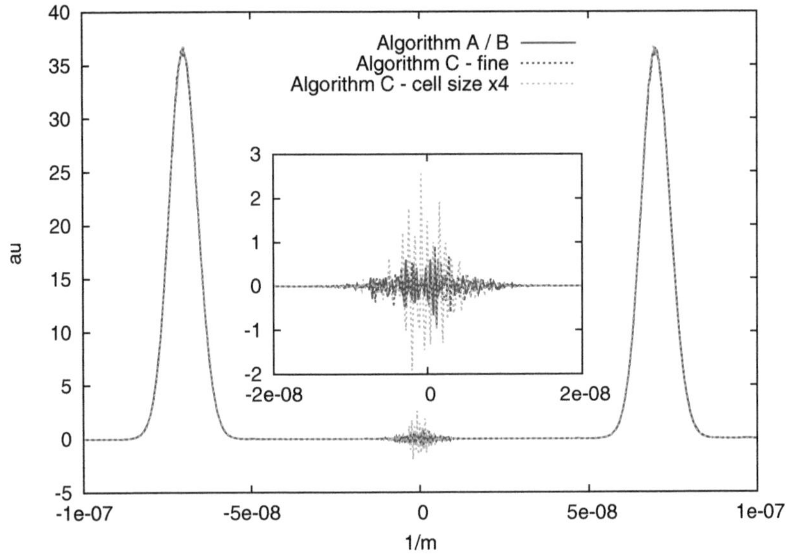

Figure 22.6. Densities $n(x) = \int d\mathbf{k} f_w(x, \mathbf{k})$ for different cell sizes after a single 10fs time step show spurious oscillations in the central quantum region, which is given in more detail in the central inset.

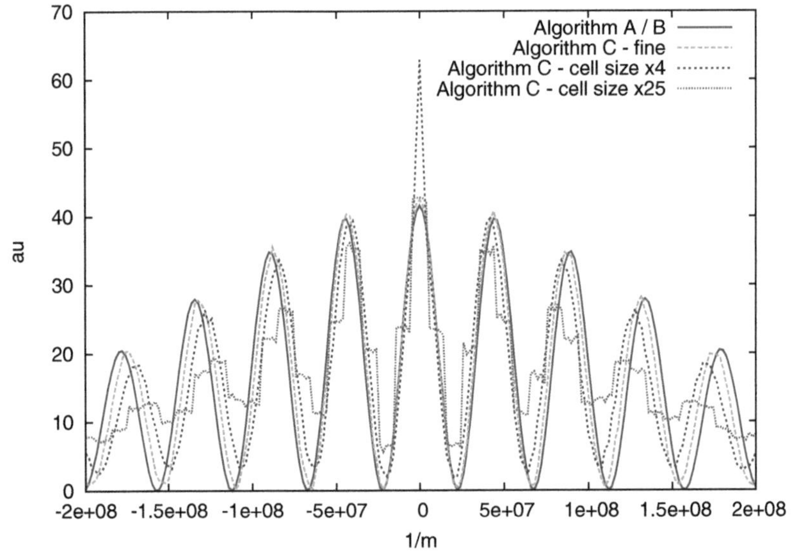

Figure 22.7. The coherent momentum distribution must remain frozen in time, as correctly shown by Algorithms A/B after 700fs evolution. The numerical scattering in Algorithm C causes decoherence similar to that of phonons with increasing cell size.

Acknowledgments. This work has been supported by the Austrian Science Fund Project FWF-P21685, the Österreichische Forschungsgemeinschaft (ÖFG) Project MOEL453, and by the Bulgarian Science Fund Project DTK 02/44.

References

[1] S. Heinrich, Quantum Complexity of Numerical Problems, in: Cucker, F. et al. (eds.), *Foundations of computational mathematics: Minneapolis 2002 (FoCM 2002)*, selected papers based on the plenary talks presented at FoCM 2002, Minneapolis, MN, USA, August 5–14, 2002. pp. 76–95, London Mathematical Society Lecture Note Series 312, Cambridge University Press, 2004.

[2] W. H. Zurek, Decoherence, Einselection, and the Quantum Origins of the Classical, *Review of Modern Physics* **75** (2003, 715–775.

[3] E. Joos et al., *Decoherence and Appearance of the Classical World in the Quantum Theory*, Springer-Verlag, New York Heidelberg Berlin, 2003.

[4] D. Querlioz and P. Dollfus, *The Wigner Monte Carlo Method for Nanoelectronic Devices*, Wiley, New York, 2010.

[5] P. Schwaha, M. Nedjalkov, S. Selberher, and I. Dimov, Phonon-Induced Decoherence in Electron Evolution, in: I. Lirkov, S. Margenov, and J. Wasniewski (eds.): *LSSC 2011*, pp. 472-479, LNCS 7116, Springer-Verlag, Berlin Heidelberg, 2012.

[6] M. Nedjalkov et al., Unified Particle Approach to Wigner-Boltzmann Transport in Small Semiconductor Devices, *Physical Review B* **70** (2004), 115319–115335.

Author information

Philipp Schwaha, Shenteq s.r.o., Bratislava, Slovak Republic.
Email: `schwaha@shenteq.com`

Mihail Nedjalkov, Vienna University of Technology, Institute for Microelectronics, Vienna, Austria.
Email: `nedjalkov@iue.tuwien.ac.at`

Siegfried Selberherr, Vienna University of Technology, Institute for Microelectronics, Vienna, Austria.
Email: `selberherr@iue.tuwien.ac.at`

Ivan Dimov, Institute of Information and Communication Technologies, Bulgarian Academy of Sciences, Sofia, Bulgaria.
Email: `ivdimov@bas.bg`

Chapter 23

Geometric Allocation Approach for the Transition Kernel of a Markov Chain

Hidemaro Suwa and Synge Todo

Abstract. We introduce a new geometric approach that constructs a transition kernel of a Markov chain. Our method always minimizes the average rejection rate and even reduces it to zero in many relevant cases, which cannot be achieved by conventional methods such as the Metropolis–Hastings algorithm or the heat bath algorithm (Gibbs sampler). Moreover, the geometric approach makes it possible to find not only a reversible but also an irreversible solution of rejection-free transition probabilities. This is the first versatile method able to construct an irreversible transition kernel in general cases. We demonstrate that the autocorrelation time (asymptotic variance) of the 8-state Potts model becomes more than 14 times as short as that with the conventional Metropolis–Hastings algorithm. Our algorithms are applicable to almost all kinds of Markov chain Monte Carlo methods and will improve efficiency.

Keywords. Markov Chain, Transition Kernel, Geometric Allocation, Detailed Balance, Reversibility.

Mathematics Subject Classification 2010. 60J10, 60J22, 65K05.

23.1 Introduction

The Markov chain Monte Carlo (MCMC) method, which is based on the importance sampling and is a powerful tool especially for systems with multiple degrees of freedom, is being applied extensively across the various disciplines, such as statistics, physics, chemistry, bioinformatics, economics, etc. [10, 16]. Although an MCMC method which satisfies appropriate conditions (ergodicity) guarantees that estimators asymptotically in principle converge [13], rapid convergence is essential for the method to work in practice. In the Monte Carlo method, if the central limit theorem holds, the variance of expectations decreases as σ^2/n, where n is the number of samples. Then, our concern is to reduce the asymptotic variance σ^2. Since the autocorrelation of a Markov chain exactly corresponds to the asymptotic variance, it is clearly important to develop an update method that achieves a shorter autocorrelation time.

There are three key points for the MCMC method to be effective. One is the choice of the ensemble. From the view of this point, the extended ensemble methods, such as the multicanonical method [4] and the replica exchange method [9], have been proposed and applied successfully to protein-folding problems, spin glasses, etc. The second point is the selection of candidate configurations. The cluster algorithms, e.g., the Swendsen–Wang algorithm [18] and the loop algorithm [6], can overcome the critical slowing down by taking advantage of mapping to graph configurations in many physical models. The third point is the determination of the transition probability, given candidate configurations. We focus our interest on this optimization problem of the probabilities throughout this chapter.

In the MCMC method, the (total) balance, that is, the invariance of target distribution, is usually imposed to the transition kernel although a kind of adaptive procedure has drawn much attention at present [1]. For the optimization of the transition probabilities, it is a guiding principle for minimizing the rejection rate, the probability that a configuration is still remaining at the previous state [14]. In most practical implementations, the Metropolis–Hastings algorithm [8, 12] (herein referred to simply as Metropolis algorithm) or the heat bath algorithm [2], namely, the Gibbs sampler [7], have been used to determine transition probabilities. These canonical algorithms satisfy the detailed balance, the reversibility, which is a sufficient condition for the total balance. Under this condition, thanks to the simple property that every elementary transition balances with a corresponding inverse process, it becomes easy to find a qualified transition probability by solving the equation for each pair of configurations. Thus, attempts to reduce autocorrelation in the optimization problem have o far concentrated within this sufficient condition [11,15]. However, all the previous methods fail to minimize the rejection rate in most cases.

In this chaper we introduce a new method, which constructs a transition kernel by a geometric approach. This method can find solutions by applying a graphical procedure, *weight allocation*, instead of solving the detailed balance equation algebraically as previously. Surprisingly, it is *always* possible to find a solution which minimizes the average rejection rate. In the meantime, it has long been considered difficult to satisfy the total balance without imposing the detailed balance. This reversibility is sufficient but *not necessary* for the invariance of target distribution. If it is possible to find a solution beyond the sufficient condition, further optimization can be achieved. Our approach is the first method able to generally satisfy the total balance without the detailed balance. We will introduce our geometric picture for the optimization problem and then explain concrete algorithms for constructing a reversible and an irreversible kernel [17]. We will demonstrate its effectiveness in a basic physical example, using the single spin update of the ferromagnetic Potts model.

23.2 Geometric Approach

In the MCMC method, we update configuration (or state) variables locally and run over the whole system. Now, let us consider updating one discrete variable as an elementary process, e. g., flipping a single spin in the Ising or Potts models [19]. Given an environmental configuration, we would have n candidates (including the current one) for the next configuration. The weight of each candidate configuration (or state) is given by w_i $(i = 1, ..., n)$, to which the equilibrium probability measure is proportional. Although the total and detailed balance are usually expressed in terms of the weights $\{w_i\}$ and the transition probabilities $\{p_{i\to j}\}$ from state i to j, it is more convenient to introduce a quantity $v_{ij} := w_i p_{i\to j}$, which corresponds to the amount of (raw) stochastic flow from state i to j. The law of probability conservation and the total balance are then expressed as

$$w_i = \sum_{j=1}^{n} v_{ij} \qquad \forall\, i \tag{23.1}$$

$$w_j = \sum_{i=1}^{n} v_{ij} \qquad \forall\, j, \tag{23.2}$$

respectively. The average rejection rate is written as $\sum_i v_{ii} / \sum_i w_i$. Also, it is straightforward to confirm that $\{v_{ij}\}$ satisfy $v_{ij} = \min[w_i, w_j]/(n-1)$ $(i \neq j)$ for the Metropolis algorithm with the flat proposal distribution, and $v_{ij} = w_i w_j / \sum_{k=1}^{n} w_k$ $(\forall\, i,\, j)$ for the heat bath algorithm (Gibbs sampler), where the detailed balance, i.e., the absence of net stochastic flow, is manifested by the symmetry under the interchange of the indices:

$$v_{ij} = v_{ji} \qquad \forall\, i,\, j. \tag{23.3}$$

Our task is to find a set $\{v_{ij}\}$ that minimizes the average rejection rate while satisfying Equations (23.1) and (23.2). The procedure for the task can be visually understood as *weight allocation*, where we move (or allocate) some amount of weight (v_{ij}) from state i to j, keeping the entire shape of the weight boxes intact. To understand this allocation picture, let us first think of the case with $n = 2$ as in the single spin update of the Ising model. Figure 23.1 shows the allocation when the Metropolis and heat bath algorithms are applied, where the average rejection rate ($\propto v_{11} + v_{22}$) clearly remains finite. Indeed, for $n = 2$ the Metropolis algorithm gives the best solution, i.e., the minimum average rejection rate even within the total balance (see Equation (23.4) below).

For $n \geq 3$, these two methods fail to minimize the rejection rate, as we will discuss below. Besides, a generic method that accomplishes the minimization has not previouly been known. We will show that we can easily make it possible by this geometric

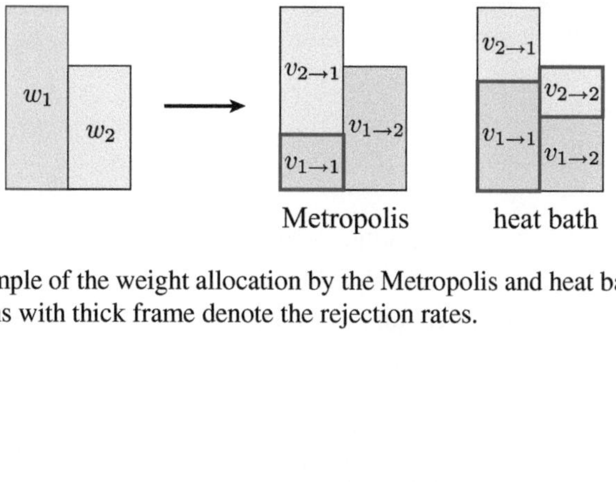

Figure 23.1. Example of the weight allocation by the Metropolis and heat bath algorithms for $n = 2$. The regions with thick frame denote the rejection rates.

Algorithm 23.1. Construction of a reversible kernel with minimized rejection.

Sort n candidate configurations as $w_1 \geq w_2 \geq w_3 \geq \cdots \geq w_n$ $(n \geq 3)$.

```
v_ij ← w_i δ_ij
w_diff ← w_1 − w_2
S_3 ← Σ_{i=3}^{n} w_i

if w_diff ≥ S_3 do
    for i = 2, ..., n do
        Swap( 1, i, w_i )                 // v_ii becomes 0
    end for
else
    for i = 3, ..., n do
        v ← w_diff * w_i / S_3
        Swap( 1, i, v )
    end for                               // v_11 = v_22 ≥ v_33 ≥ ... ≥ v_nn
    for j = n, ..., 2 do
        v' ← v_jj / (j − 1)
        for k = j − 1, ..., 1 do
            Swap( j, k, v' )
        end for                           // v_11 = v_22 ≥ ··· ≥ v_{j−1,j−1} and v_jj = 0
    end for
end if
```

picture. Although many optimal solutions are actually found, here we will introduce two specific algorithms. One makes a reversible kernel, and the other makes an *irreversible* kernel without the detailed balance.

23.2.1 Reversible Kernel

For describing our algorithm, let us introduce an operation named Swap:

```
Swap( i, j, w ) {
    v_ii ← v_ii − w
    v_ij ← v_ij + w
    v_ji ← v_ji + w
    v_jj ← v_jj − w
}.
```

We note that if $\{v_{ij}\}$ satisfy the conditions (23.1), (23.2), and (23.3), this Swap operation does not break them. A certain algorithm for the construction of reversible kernel that minimizes the average rejection rate is described in Algorithm 23.1.

This algorithm starts with the diagonal matrix $[v_{ij}]$ and uses only the Swap operation for construction. Therefore the three conditions (23.1), (23.2), and (23.3) are automatically satisfied in the whole procedure. This algorithm can be depicted visually as Algorithm 1 in Figure 23.2. As a result, the self-allocated weight that produces rejection is expressed as

$$v_{ii} = \begin{cases} \max(0,\ w_1 - \sum_{i=2}^{n} w_i) & i = 1 \\ 0 & i \geq 2. \end{cases} \tag{23.4}$$

That is, a rejection-free solution can be obtained if

$$w_1 \leq \frac{S_n}{2} \equiv \frac{1}{2} \sum_{k=1}^{n} w_k \tag{23.5}$$

is satisfied. In contrast, when inequality (23.5) is not satisfied, one must assign the maximum weight to itself, since it is larger than the sum of the rest. Thus, the present solution is optimal in the sense that it minimizes the average rejection rate.

23.2.2 Irreversible Kernel

Next, we show another algorithm that constructs an irreversible kernel [17]. The whole algorithm is described in Algorithm 23.2. In the algorithm, if two or more configurations have the same maximum weight, choose one of them first. Any order of configurations accomplishes the same minimized rejection rate. In the above procedure, all the boxes are filled without any space, as well as the reversible case, as in Figure 23.2; it satisfies the two conditions (23.1) and (23.2). However, the reversibility (23.3) is broken (for example, $v_{12} > 0$, but $v_{21} = 0$ as depicted in the figure). Since w_1 is the maximum, it is also clear that the second and subsequent boxes must already be saturated when the allocation of its own weight is initiated.

The rejection rate becomes the same with the previous reversible kernel as formulated in (23.4). In contrast to the reversible case, a net stochastic flow is introduced as the result of breaking the detailed balance, and it is expected to further boost the sampling efficiency [5].

23.3 Benchmark Test

In order to assess the effectiveness of the present algorithms, we investigate the autocorrelations in the ferromagnetic q-state Potts models on the square lattice [19], which exhibit a continuous ($q \leq 4$) or first-order ($q > 4$) phase transition at $T = 1/\ln(1 + \sqrt{q})$. We calculate the autocorrelation time of the square of order parameter for $q = 4$ and 8 by several algorithms. The autocorrelation time τ_{int} is estimated through the relation: $\sigma^2 = (1 + 2\tau_{\text{int}})\sigma_0^2$, where σ_0^2 and σ^2 are the variances of the estimator without considering autocorrelation and with calculating correlation from the binned

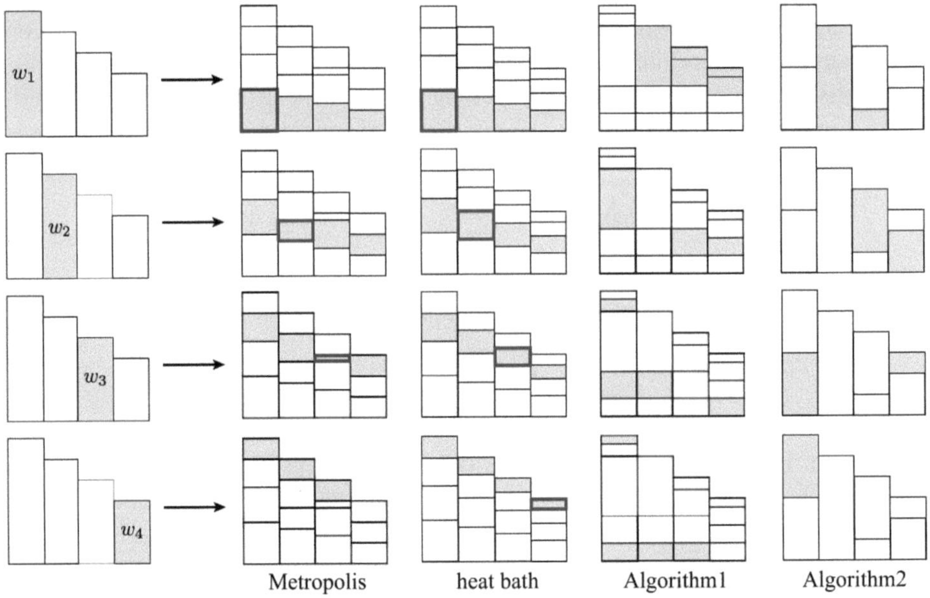

Figure 23.2. Example of weight allocation by the Metropolis, the heat bath, and the proposed two algorithms for $n = 4$. Algorithm 1 constructs a reversible kernel, and Algorithm 2 does an irreversible kernel. Both proposed algorithms minimize the average rejection rate in general, and they are rejection free in this case while the conventional methods remain finite rejection rates as indicated by the thick frames.

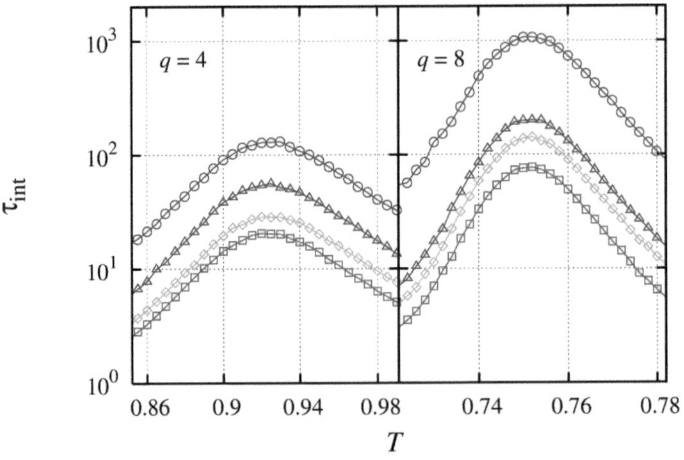

Figure 23.3. Autocorrelation time of the square of order parameter near the transition temperature ($T \simeq 0.910$ and 0.745, respectively) in the 4-state (left) and 8-state (right) Potts models by the Metropolis (circles), heat bath (triangles), LOU (diamonds), and present (squares) methods. The results of present two algorithms are the same in this scale. The system size is 16×16. The error bars are of the same order as the point sizes.

Algorithm 23.2. Construction of an irreversible kernel with minimized rejection.

```
Choose a configuration that has the maximum weight, and number it 1.
Sort other configurations in an arbitrary order.
i ← 1 j ← 2
while i ≤ n do
  w_r ← w_i
    while w_r > 0 do
      if w_r ≥ w_j then
        v_ij ← w_j
        w_r ← w_r − w_j
      if j = n then
        j ← 1
      else
        j ← j + 1
      end if
      else
        v_ij ← w_r
        w_j ← w_j − w_r
        w_r ← 0
      end if
    end while
  i ← i + 1
end while
```

data using a bin size much larger than the τ_{int} [10]. In Figure 23.3, it is clearly seen that our algorithms significantly boosts the convergence in both models in comparison with the conventional methods. In the 4 (8)-state Potts model, the autocorrelation time becomes nearly 6.4 (14) times as short as that by the Metropolis algorithm, 2.7 (2.6) times as short as the heat bath algorithm, and even 1.4 (1.8) times as short as the locally optimal update (LOU) [15], which was considered to be one of the best solutions before our approach. The autocorrelations of our two algorithms are much the same both for $q = 4, 8$. We also note that our irreversible algorithm improves the efficiency by more than 100 times compared to that of the heat bath algorithm in a quantum spin model [17].

23.4 Conclusion

We have introduced the new geometric approach for optimization of transition probabilities and the two concrete algorithms which always minimizes the average rejection rate in the MCMC method. One constructs a reversible kernel, and the other does an irreversible kernel, which is the first versatile method that constructs an irreversible chain in general cases. We showed that our algorithms significantly improve the sampling efficiency in the ferromagnetic Potts models. The autocorrelations of our two algorithms are much the same in the model; the net stochastic flow does not matter for the efficiency. However, it is generally possible for the flow to play an important role to the convergence. The introduction of efficient flow needs to be researched in the future. Finally, we note that our algorithm for irreversible kernels can be generally extended to continuous variables, which will be presented in another report.

Acknowledgments. This research is supported by Grant-in-Aid for Scientific Research Program (Nos. 20540364, 23540438) from JSPS and by Grand Challenges in Next-Generation Integrated Nanoscience, Next-Generation Supercomputer Project from MEXT, Japan. Most simulations were performed on the T2K Supercomputer at the University of Tsukuba. The program was developed based on the ALPS library [3].

References

[1] C. Andrieu and J. Toms, A tutorial on adaptive MCMC, *Stat. Comput.* **18** (2008), 343–373.

[2] A. A. Barker, Monte Carlo Calculations of The Radial Distribution Functions for A Proton-Electron Plasma, *Aust. J. Phys.*, **18** (1965), 119.

[3] B. Bauer et al., The ALPS project release 2.0: open source software for strongly correlated systems, *J. Stat. Mech.* (2011), P05001, available online at http://iopscience.iop.org/1742-5468/2011/05/P05001/ (last accessed September 21, 2012).

[4] B. A. Berg and T. Neuhaus, Multicanonical Ensemble: A New Approach to Simulate First-Order Phase Transitions, Phys. Rev. Lett. **68** (1992), 9.

[5] P. Diaconis, S. Holmes, and R. M. Neal, Analysis of a Nonreversible Markov Chain Sampler, *Ann. Appl. Probab.* **10** (2000), 726.

[6] H. G. Evertz, G. Lana and M. Marcu, Cluster algorithm for vertex models, *Phys. Rev. Lett.* **70** (1993), 875.

[7] S. Geman and D. Geman, Stochastic relaxation, Gibbs distributions and the Bayesian restoration of images, *IEEE Trans. Pattn. Anal. Mach. Intel.* **6** (1984), 721.

[8] W. K. Hastings, Monte Carlo sampling methods using Markov chains and their applications, *Biometrika* **57** (1970), 97.

[9] K. Hukushima and K. Nemoto, Exchange Monte Carlo Method and Application to Spin Glass Simulations, *J. Phys. Soc. Jpn.* **65** (1996), 1604.

[10] D. P. Landau and K. Binder, *A Guide to Monte Carlo Simulations in Statistical Physics*, 2nd ed., Cambridge University Press, Cambridge, 2005.

[11] J. S. Liu, Metropolized independent sampling with comparisons to rejection sampling and importance sampling, *Stat. Comput.* **6** (1996), p. 113.

[12] N. Metropolis, A. W. Rosenbluth, M. N. Rosenbluth, A. H. Teller, and E. Teller, Equation of State Calculations by Fast Computing Machines, *J. Chem. Phys.* **21** (1953), 1087.

[13] S. P. Meyn and R. L. Tweedie, *Markov Chains and Stochastic Stability*, Springer, New York, 1993.

[14] P. H. Peskun, Optimum Monte Carlo sampling using Markov chains, *Biometrika* **60** (1973), 607.

[15] L. Pollet, S. M. A. Rombouts, K. Van Houcke, and K. Heyde, Optimal Monte Carlo updating, *Phys. Rev. E* **70** (2004), 056705.

[16] C. P. Robert and G. Casella, *Monte Carlo Statistical Methods*, 2nd ed., Springer, New York, 2004.

[17] H. Suwa and S. Todo, Markov Chain Monte Carlo Method without Detailed Balance, *Phys. Rev. Lett.* **105** (2010), 120603.

[18] R. H. Swendsen and J. S. Wang, Nonuniversal critical dynamics in Monte Carlo simulations, *Phys. Rev. Lett.* **58** (1987), 86.

[19] F. Y. Wu, The Potts model, *Rev. Mod. Phys.* **54** (1982), 235.

Author information

Hidemaro Suwa, Department of Applied Physics, University of Tokyo, Tokyo, Japan.
Email: suwamaro@looper.t.u-tokyo.ac.jp

Synge Todo, Institute for Solid State Physics, University of Tokyo, Kashiwa, Japan.
Email: wistaria@issp.u-tokyo.ac.jp

Chapter 24

Exact Sampling for the Ising Model at All Temperatures

Mario Ullrich

Abstract. The Ising model is often referred to as the most studied model of statistical physics. It describes the behavior of ferromagnetic material at different temperatures. It is also an interesting model for mathematicians, because although the Gibbs distribution is continuous in the temperature parameter, the behavior of the usual single-spin dynamics to sample from this measure varies extremely. Namely, there is a critical temperature where we get rapid mixing above and slow mixing below this value. In this chapter we give a survey of the known results on mixing time of Glauber dynamics for the Ising model on the square lattice and present a technique that makes exact sampling of the Ising model at all temperatures possible in polynomial time. At high temperatures this is well known, but we have not found any reference where exact sampling for the Ising model at low temperatures in polynomial time is described.

Keywords. Ising Model, Exact Sampling, Random Cluster Model.

Mathematics Subject Classification 2010. 60J10, 82C44, 65C40.

24.1 Introduction

In this chapter we summarize the known results on the mixing time of the heat bath dynamics for the Ising model and combine them with some graph theoretic results to an algorithm for exact sampling from the Ising model in polynomial time. By time (or running time) we always mean the number of steps of the underlying Markov chain. The algorithm that will be analyzed (Algorithm 24.2, given in Section 24.5) is at high temperatures simply the Coupling from the past algorithm (see Propp and Wilson [18]). At low temperatures we have to produce a sample at the dual graph, but this can be traced back to sampling on the initial graph with constant boundary condition.

The main theorem of this article is stated as follows.

Theorem 24.1. *Let G_L be the square lattice with $N = L^2$ vertices. Then, Algorithm 24.2 outputs an exactly distributed Ising configuration with respect to $\pi_\beta^{G_L}$ in expected time smaller than*

- $c_\beta\, N\, (\log N)^2$ *for* $\beta \neq \beta_c = \log(1+\sqrt{2})$ *and some* $c_\beta > 0$
- $16\, N^C \log N$ *for* $\beta = \beta_c$ *and some* $C > 0$.

As a consequence we get that one can estimate the expectation of arbitrary functions with respect to the Gibbs distribution in polynomial time. Namely, if we use the simple Monte Carlo method to approximate the expectation of a function f on the Ising model, we need $\epsilon^{-2}\|f - \mathbb{E}_{\pi_\beta} f\|_2^2$ exact samples from π_β (i. e., Algorithm 24.2) to reach a mean square error of at most ϵ^2. Therefore, if we denote the bounds from Theorem 24.9 by T_β, we need on average $T_\beta\, \epsilon^{-2}\|f - \mathbb{E}_{\pi_\beta} f\|_2^2$ steps of the Markov chain that will be defined in Section 24.2.

The first polynomial-time algorithm (FPRAS) was shown by Jerrum and Sinclair [10]. There they present an algorithm to approximate the partition function Z_β of the Ising model on arbitrary graphs and, as a consequence, approximate expectations of functions that are given in terms of the partition function in polynomial time at all temperatures β.

24.2 The Ising Model

In this section we introduce the two-dimensional Ising model. Let $G = (V, E)$ be a graph with finite vertex set $V \subset \mathbb{Z}^2$ and edge set $E = \{\{u, v\} \in \binom{V}{2} : |u - v| = 1\}$, where $\binom{V}{2}$ is the set of all subsets of V with 2 elements. From now, $N := |V|$. We are interested in the square lattice, i. e., $V = \{1, \dots, L\}^2$ for some $L = \sqrt{N} \in \mathbb{N}$, because it is the most widely used case. We denote the induced graph by G_L.

The *Ising model* on G_L is now defined as the set of possible configurations $\Omega_{\mathrm{IS}} = \{-1, 1\}^V$, where $\sigma \in \Omega_{\mathrm{IS}}$ is an assignment of -1 or 1 to each vertex in V, together with the probability measure

$$\pi_\beta(\sigma) := \pi_\beta^{G_L}(\sigma) = \frac{1}{Z_\beta} \exp\Bigg\{\beta \sum_{u,v:\, u \leftrightarrow v} 1\big(\sigma(u) = \sigma(v)\big)\Bigg\},$$

where $u \leftrightarrow v$ means u and v are neighbors in G_L, Z is the normalization constant and $\beta \geq 0$ is the inverse temperature. This measure is called the Gibbs distribution with free boundary condition.

Additionally we need the notion of *boundary conditions*, but we restrict ourself here to the "all plus" and "all minus" case.

Let $V^c = \mathbb{Z}^2 \setminus V$. Then we denote the lattice G_L together with the probability measure

$$\pi_\beta^{\pm}(\sigma) := \pi_\beta^{G_L,\pm}(\sigma) = \frac{1}{\tilde{Z}_\beta} \pi_\beta^{G_L}(\sigma) \cdot \exp\Bigg\{\beta \sum_{\substack{v \in V,\, u \in V^c: \\ u \leftrightarrow v}} 1\big(\sigma(v) = \pm 1\big)\Bigg\}$$

by the Ising model with plus/minus boundary condition, respectively. One can imagine that this corresponds to the Ising model on G_L with a strip of fixed spins around, so every vertex in G_L has the same number of neighbors.

In 1944 Onsager [17] proved that there is a phase transition at $\beta = \beta_c := \ln(1+\sqrt{2})$ in the case where $V = \mathbb{Z}^2$, and we will see that this value is also important for finite lattices. Namely, the dynamics defined below is rapidly mixing if and only if $\beta \le \beta_c$.

We will use the so-called *heat bath dynamics*. These dynamics define a irreducible, aperiodic and reversible Markov chain $X^\beta = (X_i^\beta)_{i\in\mathbb{N}}$ with stationary distribution π_β by the transition matrix

$$P(\sigma, \sigma^{v,\xi}) = \frac{1}{N}\left(1 + \frac{\pi_\beta(\sigma)}{\pi_\beta(\sigma^{v,\xi})}\right)^{-1} \qquad \sigma \in \Omega_{\mathrm{IS}}, \quad v \in V,$$

where $\sigma^{v,\xi}$ with $\xi \in \{-1, 1\}$ is defined by $\sigma^{v,\xi}(v) = \xi$ and $\sigma^{v,\xi}(u) = \sigma(u), u \neq v$. The interpretation of this algorithm is very simple. In each step, choose a random $v \in V$ and assign a new value to v according to π_β conditioned on all neighbors of v. Note that the results of this chapter hold in general for all Glauber dynamics as defined in [6] which admit a monotone Coupling (see Section 24.3). For a general introduction to Markov chains see, e. g., [11], or [13] in the context of spin systems.

In the following we want to estimate how fast such a Markov chain converges to its stationary distribution. Therefore we first introduce the *total variation distance* to measure the distance between two probability measures ν and π, which is defined by

$$\|\nu - \pi\|_{\mathrm{TV}} = \frac{1}{2} \sum_{\sigma\in\Omega_{\mathrm{IS}}} |\nu(\sigma) - \pi(\sigma)|\,.$$

Now we can define the *mixing time* of the Markov chain with transition matrix P and stationary distribution π_β by

$$\tau_\beta = \min\left\{n : \max_{\sigma\in\Omega_{\mathrm{IS}}} \|P^n(\sigma,\cdot) - \pi_\beta(\cdot)\|_{\mathrm{TV}} \le \frac{1}{2\mathrm{e}}\right\}.$$

This is the expected time the Markov chain needs to get close to its stationary distribution. In fact, one can bound the spectral gap of the transition matrix P in either direction in terms of the mixing time (see, e. g., [11, Theorems 12.3 and 12.4]), so one can bound the error of a MCMC algorithm to integrate functions over Ω_{IS}, as one can read in [19]. Furthermore, if the Markov chain is rapidly mixing (i. e., the mixing time is bounded by a polynomial in the logarithm of size of the state space Ω_{IS}) we get that the problem of integration (with an unnormalized density) on the Ising model is *tractable* (see also [16]). Unfortunately there is no Markov chain that is proven to be rapidly mixing at all temperatures.

However, in this chapter we are interested in sampling exactly from the stationary distribution, but first we present the known mixing-time results for the Glauber dynamics for the Ising model. For proofs or further details we refer to the particular articles or the survey of Martinelli [13]. Of course, we can only give a small selection of references, because there are many papers leading to the results given below.

Theorem 24.2 ([14]). *Let $\beta < \beta_c$. Then there exists a constant $c_\beta > 0$ such that the mixing time of the Glauber dynamics for the Ising model with arbitrary boundary condition on G_L satisfies*

$$\tau_\beta \leq c_\beta \, N \log N.$$

Theorem 24.3 ([2]). *Let $\beta > \beta_c$. Then there exists a constant $c_\beta > 0$ such that the mixing time of the Glauber dynamics for the Ising model on G_L satisfies*

$$\tau_\beta \geq e^{c_\beta L}.$$

The results above can be obtained by observing that some spatial mixing property of the measure π_β is equivalent to the mixing in time of the Glauber dynamics. For details for this interesting fact, see [3].

The constant c_β of Theorem 24.2 is widely believed to diverge if β tends to β_c. Determining the mixing time in the case $\beta = \beta_c$ was a challenging problem for a long time. It was solved by Lubetzky and Sly in their recent paper [12].

Theorem 24.4 ([12]). *There exists a constant $C > 0$ such that the mixing time of the Glauber dynamics for the Ising model on G_L at the critical temperature satisfies*

$$\tau_{\beta_c} \leq 4\, N^C.$$

Remark 24.5. We give here only a brief description of the constant C, which can be given explicitly. For more details see [12, p.19].

However, numerical experiments on the "true" exponent suggest that $C \approx 3.08$ (see, e. g., [15, 20] and note the explanation below).

The constant C in Theorem 24.4 is given by

$$C = 2 + \log_{3/2}\left(\frac{2}{1 - p^+}\right). \tag{24.1}$$

Here, p^+ is the limiting vertical crossing probability in the random cluster model on a fully-wired rectangle, where the width of the lattice is $3\times$ its height. The C, as given here, differs from the one given in [12] by eliminating a factor of 2 in front of the log term and by the additional 2. The reason is that we state their result in terms of N and not in the side length L of the lattice (therefore without factor 2) and that we are interested in the discrete time single-spin algorithms. Therefore we get an additional factor N in their spectral gap result ([12, Thm. 1]) and a factor N by (see, e. g., [11])

$$\tau_{\beta_c} \le \log\left(\frac{e}{\min_\sigma \pi_{\beta_c}(\sigma)}\right) \mathbf{gap}(X^{\beta_c})^{-1} \le 4N\,\mathbf{gap}(X^{\beta_c})^{-1},$$

because $\min_\sigma \pi_{\beta_c}(\sigma) \ge \exp(-3N)$.

The results of this section show that the Glauber dynamics is mixing rapidly for $\beta \le \beta_c$, but very slowly mixing for larger β. In Section 24.4 we will see how to avoid this problem.

24.3 Exact Sampling

In this section we briefly describe the so-called *Coupling from the past algorithm* (CFTP) to sample exactly from the stationary distribution of a Markov chain.

This algorithm works under weak assumptions on the Markov chain for every finite state space and every distribution, but to guarantee that the algorithm is efficient we need some monotonicity property of the model and that the chain is rapidly mixing. For a detailed description of CFTP and the proof of correctness see [18].

We restrict ourself to the heat bath dynamics for the Ising model. First note that the heat bath dynamics, as defined above, admits a monotone Coupling, that is, given two realizations of the heat bath chain $X = (X_t)_{t\in\mathbb{N}}$ and $Y = (Y_t)_{t\in\mathbb{N}}$, there exists a Coupling (X, Y) (i. e., using the same random numbers) such that

$$X_t \le Y_t \implies X_{t+1} \le Y_{t+1} \qquad \text{for all } t \in \mathbb{N},$$

where $\le$ means smaller or equal at each vertex.

Additionally we know that $-\mathbf{1} \le \sigma \le \mathbf{1}$ for all $\sigma \in \Omega_{\mathrm{IS}}$, where $-\mathbf{1} = (-1)^V$ and $\mathbf{1} = (1)^V$. Therefore if we set $X_0 = -\mathbf{1}$ and $Y_0 = \mathbf{1}$ we know that $X_0 \le \sigma \le Y_0$ for all σ and so $X_t \le Z_t \le Y_t$ for the realization $Z = (Z_t)_{t\in\mathbb{N}}$ with $Z_0 = \sigma$. Since this holds for all σ, one can choose $Z_0 \sim \pi_\beta$ and we get that whenever X_t and Y_t coalesce, they also coalesce with Z_t which has the right distribution.

Having presented the idea of the algorithm, we will now state the algorithm in detail. Note that the algorithm is called Coupling from the past, because we run the chains from the past to the present. The algorithm CFTP(G, β) to sample from the distribution π_β^G works as described in Algorithm 1.

We denote the algorithm by CFTP$^\pm(G, \beta)$ if we sample with respect to $\pi_\beta^\pm$, i. e., with plus/minus boundary condition.

See [9] for examples that show that it is necessary to go from the past in the future and that we have to reuse the random numbers.

Now we state the connection between the expected running time of the CFTP algorithm and the mixing time of the Markov chain.

Algorithm 24.1. Coupling from the past.

Input: The graph $G = (V, E)$ and the value of β
Output: An Ising configuration $\sigma \sim \pi_\beta$

```
procedureCFTPG, β
    Set t = 0
    Set X_0 = −1 and Y_0 = 1
    while X_0 ≠ Y_0
        t = t + 1
        Generate random numbers U_{−2^t+1}, ..., U_{−2^{t−1}} that are sufficient to run
            the Markov chain. (e. g., U_i ∼ Uniform {V × [0, 1]})
        Set X_{−2^t+1} = 0 and Y_{−2^t+1} = 1 and run the chains until time 0 by using
            only the random numbers U_{−2^t+1}, ..., U_{−1}
    end while
    return σ = X_0
end procedure
```

Proposition 24.6 ([18]). *Let T_β be the expected running time of CFTP(G, β) from Algorithm 24.1 with $G = (V, E)$ and $|V| = N$. Then*

$$T_\beta \;\le\; 4\,\tau_\beta\,\log N,$$

where τ_β is the mixing time of the underlying Markov chain.

We see that exact sampling from the Gibbs distribution is efficient whenever the Markov chain is rapidly mixing. By the results of Section 24.2 we know that this is the case for $\beta \le \beta_c$. In the case $\beta > \beta_c$ we need a different technique to generate exact samples. Therefore we need essentially the so called random cluster model, as we will see in the next section.

24.4 The Random Cluster Model

The *random cluster model* (also known as the FK model) was introduced by Fortuin and Kasteleyn in [5] to study lattice spin systems with a graph structure. It is defined on a graph $G = (V, E)$ by its state space $\Omega_{\mathrm{RC}} = \{\omega : \omega \subseteq E\}$ and the RC measure

$$\mu_p(\omega) \;=\; \frac{1}{Z}\, p^{|\omega|}\,(1-p)^{|E|-|\omega|}\,2^{C(\omega)},$$

where $p \in (0, 1)$, Z is the normalization constant and $C(\omega)$ is the number of connected components in the graph (V, ω). For a detailed introduction and related topics see [7].

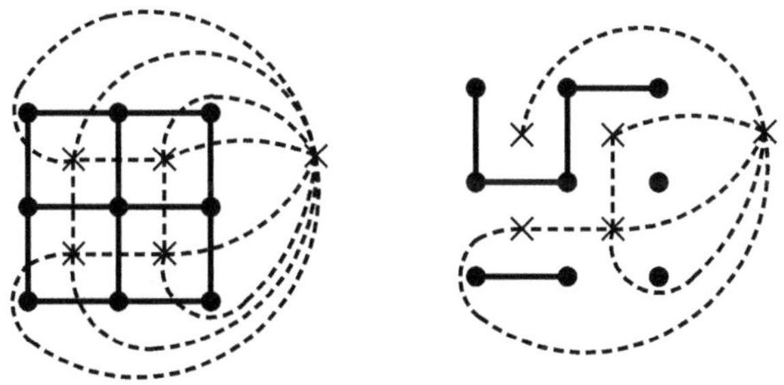

Figure 24.1. Left: The graph G_3 (solid lines) and its dual (dashed lines). Right: A RC state on G_3 (solid lines) and its dual configuration (dashed lines).

There is a tight connection between the Ising model and the random cluster model. Namely, if we set $p = 1 - e^{-\beta}$, we can translate an Ising configuration $\sigma \sim \pi_\beta$ to a random cluster state $\omega \sim \mu_p$ and vice versa. To get an Ising configuration $\sigma \in \Omega_{\mathrm{IS}}$ from $\omega \in \Omega_{\mathrm{RC}}$ assign independent and uniformly random spins to each connected component of ω. For the reverse way include all edges $e = \{e_1, e_2\} \in E$ with $\sigma(e_1) = \sigma(e_2)$ to ω with probability p. For details see [4].

Therefore sampling an Ising configuration according to π_β is equivalent to sampling a RC state from μ_p whenever both models are defined on the same graph G and $p = 1 - e^{-\beta}$.

Another important concept in connection with the RC model is the duality of graphs (see, e. g., [8]). Let $G = (V, E)$ be a finite, planar graph, i. e., without intersecting edges if we draw it in the plane (like our G_L). The *dual graph* $G^* = (V^*, E^*)$ of G is constructed as follows. Put a vertex in each face (including the infinite outer one) of the graph and connect two vertices by an edge if and only if the corresponding faces of G share a boundary edge. It is clear, that the number of vertices can differ in the dual graph, but we have the same number of edges.

Additionally we define a *dual configuration* $\omega^* \subseteq E^*$ in G^* to a RC state $\omega \subseteq E$ in G by

$$e \in \omega \iff e^* \notin \omega^*,$$

where e^* is the edge in E^* that "crosses" e. (By the construction, this edge is unique.) See Figure 24.1 for the graph G_L with $L = 3$ and its dual graph G_L^* together with 2 corresponding RC states.

Now we can state the following theorem about the relation of the distribution of a RC state and its dual (see [8]).

Proposition 24.7 ([8, p. 164]). *Let $G = (V, E)$ be a finite, planar graph and μ_p be the random cluster measure on G. Furthermore let $G^* = (V^*, E^*)$ be the dual graph of G and $\mu^*_{p^*}$ be the random cluster measure on G^*.*

Then

$$\omega \sim \mu_p \iff \omega^* \sim \mu^*_{p^*},$$

where

$$p^* = 1 - \frac{p}{2-p}. \tag{24.2}$$

Obviously, $(p^*)^* = p$. By Proposition 24.7 one can see that sampling from μ_p and sampling from $\mu^*_{p^*}$ is equivalent. It is straightforward to get the following proposition.

Proposition 24.8. *Sampling from the Gibbs distribution π^G_β is equivalent to sampling from the Gibbs distribution $\pi^{G^*}_{\beta^*}$, where*

$$\beta^* = \log\left(\coth\frac{\beta}{2}\right). \tag{24.3}$$

Additionally,

$$\beta > \beta_c \iff \beta^* < \beta_c.$$

Note that the relations of (24.2) and (24.3) can be alternatively stated as

$$0 = 2 - 2p - 2p^* + pp^*$$

and

$$0 = 1 + e^{\beta} + e^{\beta^*} + e^{\beta+\beta^*}.$$

Proof. The equivalence was shown by the above procedure, i. e., if we want to sample from π^G_β, we can sample from $\pi^{G^*}_{\beta^*}$, generate a RC state with respect to $\mu^*_{p^*}$, go to the dual lattice with measure μ_p, and finally generate a state according to π^G_β. Since $p^{(*)} = 1 - e^{-\beta^{(*)}}$, the formula for β^* comes from

$$\beta^* = -\log(1-p^*) \overset{(24.2)}{=} \log\left(\frac{2-p}{p}\right) = \log\left(\coth\frac{\beta}{2}\right).$$

This proves the statement. □

24.5 Exact Sampling for the Ising Model

In this section we present an algorithm to sample exactly from the a Gibbs distribution. But, before we prove that it is efficient, we state our sampling algorithm.

Therefore we first have to explain what the graph G^*_L looks like. It is easy to obtain (see Figure 24.1) that $G^*_L = (V^*_L, E^*_L)$ is also a square lattice with $(L-1)^2$ vertices and an additional auxiliary vertex v^*, which is connected to every vertex on the boundary of it. We denote the operation of adding a vertex to a graph and connect it to all boundary vertices by $\cup_b$. So $G^*_L = G_{L-1} \cup_b v^*$.

Algorithm 24.2. Sampling from the Ising model on the square lattice.

Input: An integer L and the value of β
Output: An Ising configuration $\sigma \sim \pi_\beta^{G_L}$

1: **if** $\beta \le \beta_c$ **then**
2: $\quad\sigma = \mathrm{CFTP}(G_L, \beta)$
3: **else**
4: $\quad\widetilde{\sigma} = \mathrm{CFTP}^+(G_{L-1}, \beta^*)$, where β^* is given in (24.3)
5: $\quad$Define a Ising configuration σ^* on $G_L^* = G_{L-1} \cup_b v^*$ by $\sigma^*(v) = \widetilde{\sigma}(v)$ on $V(G_{L-1})$ and $\sigma^*(v^*) = 1$.
6: $\quad$Generate a RC state ω^* from σ^*
7: $\quad$Take the dual RC state $\omega = (\omega^*)^*$
8: $\quad$Generate an Ising configuration σ from ω
9: **end if**
10: **return** σ

Theorem 24.9. *Let G_L be the square lattice with $N = L^2$ vertices. Then, the algorithm from above outputs an exactly distributed Ising configuration with respect to $\pi_\beta^{G_L}$ in expected time smaller than*

- $c_\beta\, N\, (\log N)^2$ $\quad$ *for $\beta \neq \beta_c = \log(1+\sqrt{2})$ and some $c_\beta > 0$;*
- $16\, N^C \log N$ $\quad$ *for $\beta = \beta_c$, where C is given in* (24.1).

Proof. The running time of the algorithm follows directly from Theorems 24.2 and 24.4 and Proposition 24.6. Therefore we only have to prove that the output σ of the algorithm has the right distribution. In the case of $\beta \le \beta_c$ this is obvious. For $\beta > \beta_c$ we know from Proposition 24.8 that $\sigma \sim \pi_\beta^{G_L}$, if the dual configuration σ^* on G_L^* (line 5 of Algorithm 24.2) is distributed according to $\pi_{\beta^*} := \pi_{\beta^*}^{G_L^*}$. But by the construction of lines 4 and 5 of Algorithm 24.2, this is true. For this, note that $\pi_\beta(\eta) = \pi_\beta(-\eta)$ for all $\eta \in \Omega_{\mathrm{IS}}$. We get that for each vertex $v \in V$ (especially for v^*)

$$\begin{aligned}
\pi_\beta(\eta) &= \pi_\beta\big(\eta \cap \{\sigma : \sigma(v) = 1\}\big) + \pi_\beta\big(\eta \cap \{\sigma : \sigma(v) = -1\}\big) \\
&= \pi_\beta\big(\{\sigma : \sigma(v) = 1\}\big)\, \pi_\beta\big(\eta \,\big|\, \{\sigma : \sigma(v) = 1\}\big) \\
&\quad + \pi_\beta\big(\{\sigma : \sigma(v) = -1\}\big)\, \pi_\beta\big(\eta \,\big|\, \{\sigma : \sigma(v) = -1\}\big) \\
&= \frac{1}{2}\Big[\pi_\beta\big(\eta \,\big|\, \{\sigma : \sigma(v) = 1\}\big) + \pi_\beta\big(\eta \,\big|\, \{\sigma : \sigma(v) = -1\}\big)\Big] \\
&= \frac{1}{2}\, \pi_\beta\big(\{\eta, -\eta\} \,\big|\, \{\sigma : \sigma(v) = 1\}\big).
\end{aligned}$$

The last equality comes from the fact that

$$\pi_\beta\big(\eta\,\big|\,\{\sigma : \sigma(v) = -1\}\big) \;=\; \pi_\beta\big(-\eta\,\big|\,\{\sigma : \sigma(v) = 1\}\big).$$

Therefore we can sample from π_β on G_L^* by sampling η from the conditional measure $\pi_\beta(\cdot \mid \{\sigma : \sigma(v^*) = 1\})$ and then choose with probability $\frac{1}{2}$ either η or $-\eta$. If we now use that $G_L^* = G_{L-1} \cup_b v^*$ one can see that sampling on G_L^* with respect to $\pi_\beta(\cdot \mid \{\sigma : \sigma(v^*) = 1\})$ is the same as sampling $\widetilde{\sigma}$ from $\pi_\beta^{G_{L-1},+}$ and setting

$$\sigma(v) \;=\; \begin{cases} \widetilde{\sigma}(v) & v \in V(G_{L-1}), \\ 1 & v = v^*. \end{cases}$$

Note that we omit the step of choosing σ or $-\sigma$ with probability $\frac{1}{2}$, because the RC state that will be generated would be the same.

This completes the proof. □

Remark 24.10. Note that the same technique works also for the q-state Potts model. This model consists of the state space $\Omega_{\mathrm{P}} = \{1, \ldots, q\}^V$ and the same measure π_β. In this case we consider the random cluster measure

$$\mu_{p,q}(\omega) \;=\; \frac{1}{Z}\, p^{|\omega|}\,(1-p)^{|E|-|\omega|}\, q^{C(\omega)}$$

and the connection of the models is again given by $p = 1 - e^{-\beta}$.

A recent result of Beffara and Duminil-Copin [1] shows that the self-dual point of the RC model corresponds to the critical temperature of the Potts model $\beta_c(q) = \ln(1 + \sqrt{q})$ in the same way as in the case $q = 2$ (i. e., the Ising case). Therefore, a sampling algorithm for the Potts model above (and at) the critical temperature is enough to sample at all temperatures.

Acknowledgments. The author was supported by the DFG GK 1523.

References

[1] V. Beffara and H. Duminil-Copin, The self-dual point of the two-dimensional random-cluster model is critical for $q \geq 1$, *Probab. Theory Related Fields* **153** (2012), 511–542.

[2] F. Cesi, G. Guadagni, F. Martinelli, and R. H. Schonmann, On the two-dimensional stochastic Ising model in the phase coexistence region near the critical point, *J. Statist. Phys.* **85** (1996), 55–102.

[3] M. Dyer, A. Sinclair, E. Vigoda, and D. Weitz, Mixing in time and space for lattice spin systems: a combinatorial view, *Random Structures Algorithms* **24** (2004), 461–479.

[4] S. F. Edwards and A. D. Sokal, Generalization of the Fortuin-Kasteleyn-Swendsen-Wang representation and Monte Carlo algorithm, *Phys. Rev. D* **38** (1988), 2009–2012.

[5] C. M. Fortuin and P. W. Kasteleyn, On the random-cluster model. I. Introduction and relation to other models, *Physica* **57** (1972), 536–564.

[6] R. J. Glauber, Time-dependent statistics of the Ising model, *J. Mathematical Phys.* **4** (1963), 294–307.

[7] G. Grimmett, *The random-cluster model*, Grundlehren der Mathematischen Wissenschaften 333, Springer, Berlin, 2006.

[8] G. Grimmett, *Probability on Graphs*, IMS Textbooks Series 1, Cambridge University Press, Cambridge, 2010.

[9] O, Häggström, *Finite Markov chains and algorithmic applications*, London Mathematical Society Student Texts 52, Cambridge University Press, Cambridge, 2002.

[10] M. Jerrum and A. Sinclair, Polynomial-time approximation algorithms for the Ising model, *SIAM J. Comput.* **22** (1993), 1087–1116.

[11] D. A. Levin, Y. Peres, and E. L. Wilmer, *Markov chains and mixing times*. American Mathematical Society, Providence, RI, 2009, With a chapter by J. G. Propp and D. B. Wilson.

[12] E. Lubetzky and A. Sly, Critical Ising on the square lattice mixes in polynomial time, *Comm. Math. Phys.* **313** (2012), 815–836.

[13] F. Martinelli, Lectures on Glauber dynamics for discrete spin models, in: *Lectures on probability theory and statistics* (Saint-Flour, 1997), pp. 93–191, Lecture Notes in Mathematics 1717, Springer, Berlin, 1999.

[14] F. Martinelli and E. Olivieri, Approach to equilibrium of Glauber dynamics in the one phase region. I. The attractive case, *Comm. Math. Phys.* **161** (1994), 447–486.

[15] M. P. Nightingale and H. W. J. Blöte, Dynamic Exponent of the Two-Dimensional Ising Model and Monte Carlo Computation of the Subdominant Eigenvalue of the Stochastic Matrix, *Physical Review Letters* **76** (1996), 4548–4551.

[16] E. Novak and H. Woźniakowski, *Tractability of multivariate problems. Volume II: Standard information for functionals*, EMS Tracts in Mathematics 12. European Mathematical Society (EMS), Zürich, 2010.

[17] L. Onsager, Crystal statistics. I. A two-dimensional model with an order-disorder transition, *Phys. Rev.* **65** (1944), 117–149.

[18] J. G. Propp and D. B. Wilson, Exact sampling with coupled Markov chains and applications to statistical mechanics, *Random Structures Algorithms* **9** (1996), 223–252.

[19] D. Rudolf, Explicit error bounds for lazy reversible Markov chain Monte Carlo, *J. Complexity* **25** (2009), 11–24.

[20] F. Wang, Naomichi Hatano and Masuo Suzuki, Study on dynamical critical exponents of the Ising model using the damage spreading method, *J. Phys. A: Math. Gen.* **28** (1995), 4543–4552.

Author information

Mario Ullrich, Friedrich Schiller University, Jena, Germany.
Email: `mario.ullrich@uni-jena.de`

10785 Berlin
productsafety@degruyterbrill.com

Bei Fragen zur Produktsicherheit wenden Sie sich bitte an:
If you have any questions regarding product safety,
please contact:

Walter de Gruyter GmbH
Genthiner Straße 13
10785 Berlin
productsafety@degruyterbrill.com